Space Science, Exploration and Policies

snova
New York

Space Science, Exploration and Policies

Commercial Human Spaceflight: Background, Regulations and Safety
Dana G. Pritchard (Editor)
ISBN: 979-8-89530-176-0
eBook ISBN: 979-8-89530-195-1

A Guide to Black Holes
Kenath Arun, PhD (Editor)
ISBN: 979-8-88697-163-7
eBook ISBN: 979-8-88697-515-4

Conceptual Features of Einstein's Theory of General Relativity Based on the Philosophy of Science
Jun-Young Oh
ISBN: 979-8-88697-107-1
eBook ISBN: 979-8-88697-140-8

Time-Out of Time: A Postscript to Nuclear Time Travel
Bernd Schmeikal, PhD
ISBN: 978-1-68507-320-6
eBook ISBN: 978-1-68507-465-4

Solar System: Structure, Formation and Exploration
Matteo de Rossi (Editor)
ISBN: 978-1-62100-057-0
eBook ISBN: 978-1-62100-081-5

Ultra High Energy Cosmic Rays: A New Tool for Astrophysics Research
Anna Uryson
ISBN: 978-1-61668-847-9
eBook ISBN: 978-1-61728-258-4

More information about this series can be found at
https://novapublishers.com/product-category/series/space-science-exploration-and-policies/

Brent M. Jenkins
Editor

The National Aeronautics and Space Administration's Role in Space Exploration and Development

snova
New York

Additional color graphics may be available in the e-book version of this book.

Library of Congress Cataloging-in-Publication Data

ISBN: 979-8-89530-539-3 (Hardcover)
ISBN: 979-8-89530-564-5 (eBook)

Published by Nova Science Publishers, Inc. † New York

Contents

Preface		vii
Chapter 1	**National Aeronautics and Space Administration (NASA): A Primer** Rachel Lindbergh	1
Chapter 2	**International Collaboration and Competition in Space: Oversight of Nasa's Roles and Programs** Committee on Commerce, Science, and Transportation	27
Chapter 3	**NASA Exploration Plans: Where We've Been and Where We're Going** Committee on Commerce, Science, and Transportation	129
Chapter 4	**NASA's Plan for Sustained Lunar Exploration and Development** National Aeronautics and Space Administration	199
Chapter 5	**Current IV&V Projects** National Aeronautics and Space Administration	217
Chapter 6	**NASA's Deep Space Exploration Plans** National Aeronauics and Space Administration	227
Chapter 7	**NASA Appropriations and Authorizations: Fact Sheet** Rachel Lindbergh	243
Index		251

Preface

The National Aeronautics and Space Administration (NASA) was established in 1958 by the National Aeronautics and Space Act to conduct civilian space and aeronautics activities. According to its 2022 strategic plan, NASA's mission is to "explore the unknown in air and space, innovate for the benefit of humanity, and inspire the world through discovery." NASA has programs in science, human spaceflight, space technology development, aeronautics research, and science, technology, engineering, and mathematics (STEM) education.

This book describes the organization of NASA, its major programs, and its budget, providing an in-depth look at how NASA continues to push the boundaries of exploration and innovation.

Chapter 1

National Aeronautics and Space Administration (NASA): A Primer*

Rachel Lindbergh

Summary

The National Aeronautics and Space Administration (NASA) was established in 1958 by the National Aeronautics and Space Act (P.L. 85-568) to conduct civilian space and aeronautics activities. According to its 2022 strategic plan, NASA's mission is to "explore the unknown in air and space, innovate for the benefit of humanity, and inspire the world through discovery." With a budget of $25.6 billion in FY2023, about 17,000 civil service employees, and about 35,000 contractors, NASA has programs in science, human spaceflight, space technology development, aeronautics research, and science, technology, engineering, and mathematics (STEM) education. The agency has nine NASA- operated centers, a tenth center operated under contract as a federally funded research and development center (FFRDC), and other facilities. It also provides funding to academia and industry to conduct research, develop systems such as new spacecraft, and provide services such as space launch.

Among the issues facing Congress—through the annual appropriations process and potentially in authorization legislation— are oversight and funding for the Artemis initiative to resume human exploration of the Moon for the first time since the Apollo program; the future of the International Space Station (ISS) and oversight of the commercial providers that transport crews and cargo to and from the ISS under NASA

---

* This is an edited, reformatted and augmented version of Congressional Research Service Publication No. R47891, dated January 3, 2024.

In: The National Aeronautics and Space Administration's Role …
Editor: Brent M. Jenkins
ISBN: 979-8-89530-539-3

contract; research priorities in Earth science, planetary science, and other fields; and aeronautics initiatives such as the development of experimental demonstrator aircraft for supersonic passenger flight and electric propulsion.

The core capabilities for the Artemis program are a crew capsule, known as Orion, and a heavy-lift rocket to launch Orion into space, known as the Space Launch System (SLS). Orion and the SLS have been in development since a congressional mandate in the NASA Authorization Act of 2010 (P.L. 111-267). An uncrewed test flight of Orion and SLS, known as Artemis I, was completed successfully in 2022. A first crewed test, known as Artemis II, is anticipated in 2024. The crewed Artemis III mission, including the first human lunar landing since 1972, is anticipated in 2025. Subsequent Artemis missions are planned.

Between 2012 and 2019, ISS crews, including U.S. astronauts, were carried exclusively by Russian Soyuz spacecraft. NASA has crew transport contracts with two U.S. companies, Space Exploration Technologies (SpaceX) and Boeing. The first operational crewed flight by SpaceX was in 2020. Boeing plans a test flight in 2024. ISS operation is authorized through at least FY2030 by the NASA Authorization Act of 2022 (P.L. 117-167, §10815). At some point, even if operations are extended further, the ISS will reach the end of its useful life. After the ISS, NASA expects to rely on commercially operated space stations to provide services and facilities in low Earth orbit (LEO). NASA's Commercial LEO Development program is supporting several companies to develop such capabilities. NASA has acknowledged that there may be a gap between the end of ISS operations and the first availability of commercial alternatives.

NASA is led by an Administrator, who is subject to confirmation by the Senate. Five mission directorates manage most of the agency's programs: Exploration Systems Development, Space Operations, Science, Space Technology, and Aeronautics Research. Exploration Systems Development and Space Operations manage human spaceflight activities, while Science manages robotic spaceflight and other scientific research. In addition, the Office of STEM Engagement manages NASA education programs. The five mission directorates and the Office of STEM Engagement each have their own appropriations account and are each led by an Associate Administrator. Separate appropriations accounts fund Construction and Environmental Compliance and Remediation; Safety, Security, and Mission Services; and the NASA Inspector General.

Introduction

The National Aeronautics and Space Administration (NASA) was established in 1958 by the National Aeronautics and Space Act (P.L. 85-568) to conduct civilian space and aeronautics activities. According to its 2022 strategic plan, NASA's mission is to "explore the unknown in air and space, innovate for the benefit of humanity, and inspire the world through discovery."[1] With a budget of $25.6 billion in FY2023, about 17,000 civil service employees, and about 35,000 contractors, NASA has programs in science, human spaceflight, space technology development, aeronautics research, and science, technology, engineering, and mathematics (STEM) education. The agency has nine NASA-operated centers, a tenth center operated under contract as a federally funded research and development center (FFRDC), and other facilities. It also provides funding to academia and industry to conduct research, develop systems such as new spacecraft, and provide services such as space launch.

Among the issues facing Congress—through the annual appropriations process and potentially in authorization legislation—are oversight and funding for the Artemis initiative to resume human exploration of the Moon for the first time since the Apollo program; the future of the International Space Station (ISS) and oversight of the commercial providers that transport crews and cargo to and from the ISS under NASA contract; research priorities in Earth science, planetary science, and other fields; and aeronautics initiatives such as the development of experimental demonstrator aircraft for supersonic passenger flight and electric propulsion.

This report describes the organization of NASA, its budget, and its major programs. It also provides brief discussions of selected other topics of frequent congressional interest. The intent is to provide an overview of NASA and its activities, along with references for further information.

Organization

NASA is led by an Administrator, who is "appointed from civilian life by the President by and with the advice and consent of the Senate."[2] The current

[1] NASA, *NASA Strategic Plan 2022*, p. 3, https://www.nasa.gov/wp-content/uploads/2023/09/fy-22-strategic-plan-1.pdf.

[2] 51 U.S.C. §20111(a). The meaning of the phrase *civilian life* was much discussed in 2009 when President Obama nominated Charles F. Bolden, Jr., a Major General in the U.S. Marine Corps who had worked in the private sector since retiring from the military in 2003.

Administrator is former Senator Bill Nelson. The NASA Administrator is not a member of the President's Cabinet. Three other NASA positions are also subject to Senate confirmation: the Deputy Administrator,[3] Chief Financial Officer,[4] and Inspector General.[5]

The organization of NASA is documented in NASA Policy Directive NPD 1000.3E.[6]

Management of most of the agency's major programs is organized into five mission directorates: Exploration Systems Development, Space Operations, Science, Space Technology, and Aeronautics Research (see Figure 1).[7] In addition, the Office of STEM Engagement manages NASA education programs. The five mission directorates and the Office of STEM Engagement each have their own appropriations account and are each led by an Associate Administrator. The programs managed by these offices are discussed in more detail later in this report.

Centers

NASA operates nine centers:

- Ames Research Center, Moffett Field, CA;
- Armstrong Flight Research Center, Edwards, CA;
- Glenn Research Center, Cleveland, OH;
- Goddard Space Flight Center, Greenbelt, MD;
- Johnson Space Center, Houston, TX;
- Kennedy Space Center, FL;
- Langley Research Center, Hampton, VA;

For more details, see the record of his nomination hearing before the Senate Commerce Committee, July 8, 2009, https://www.govinfo.gov/content/pkg/CHRG-111shrg54285/pdf/CHRG- 111shrg54285.pdf, including memoranda from the Department of Justice (p. 83) and from CRS (p. 89). General Bolden was confirmed as NASA Administrator on July 15, 2009.

[3] 51 U.S.C. §20111(b).

[4] 31 U.S.C. §901.

[5] 5 U.S.C. §403.

[6] NASA, *The NASA Organization*, NASA Policy Directive NPD 1000.3E, https://nodis3.gsfc.nasa.gov/npg_img/N_PD_1000_003E_/N_PD_1000_003E_.pdf. See also https://www.nasa.gov/organization/.

[7] Before September 21, 2021, the Exploration Systems Mission Directorate and the Space Operations Mission Directorate were combined as the Human Exploration and Operations Mission Directorate.

- Marshall Space Flight Center, Huntsville, AL; and
- Stennis Space Center, MS.

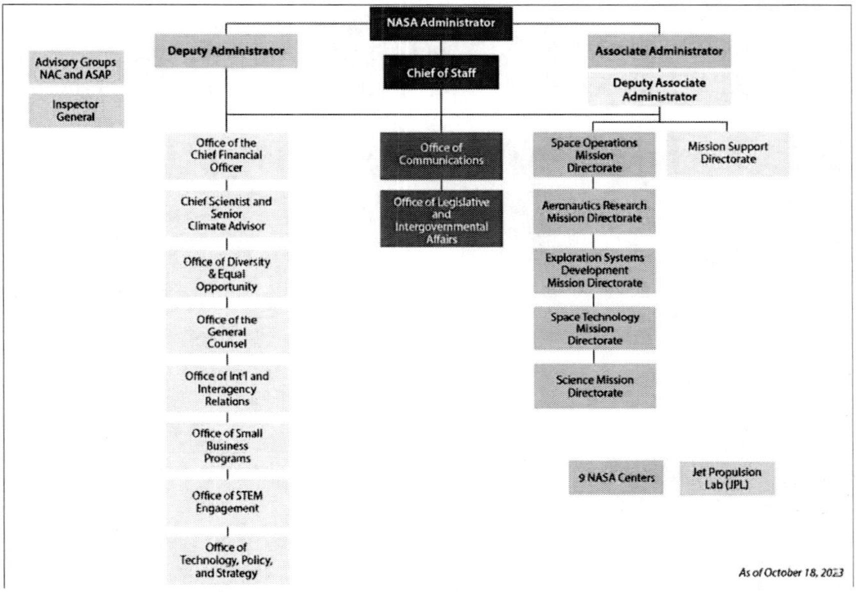

Source: CRS, adapted from https://www.nasa.gov/wp-content/uploads/2023/10/nasa-org-chart-oct-2023.pdf.

Notes: NAC = NASA Advisory Council; ASAP = Aerospace Safety Advisory Panel; STEM = science, technology, engineering, and mathematics.

Figure 1. NASA Organization Chart.

The Jet Propulsion Laboratory (JPL), in Pasadena, CA, is an FFRDC operated by the California Institute of Technology (Caltech) under a NASA contract; it is sometimes considered a tenth center.[8] Other NASA facilities are operated under the auspices of a center and do not themselves have center status. See Figure 2.

The Directors of the NASA centers report directly to the Office of the Administrator, not through the mission directorates. Congress does not appropriate funds directly to the centers, but some centers focus primarily on the programs of a particular mission directorate. For example, Goddard Space Flight Center is funded mostly by the Science Mission Directorate, while

[8] For more information on FFRDCs, see CRS Report R44629, Federally Funded Research and Development Centers (FFRDCs): Background and Issues for Congress, by Marcy E. Gallo.

Marshall Space Flight Center is funded mostly by the Exploration Systems Development Mission Directorate.

JPL reports programmatically to (and is mostly funded by) the Science Mission Directorate.

Deferred maintenance at NASA facilities is a perennial challenge.[9] As of March 2023, NASA reported a deferred maintenance backlog of $3 billion.[10]

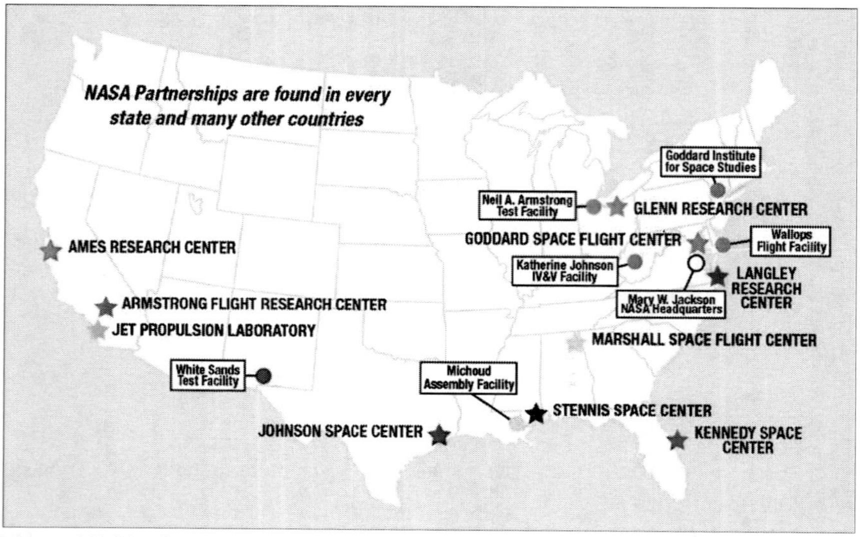

Source: NASA, https://www.nasa.gov/partnerships/nasa-locations-capabilities-and-points-of-contact/.

Notes: Alaska and Hawaii not shown. Stars indicate centers and the Jet Propulsion Laboratory, a federally funded research and development center. Circles indicate other facilities, colored to indicate the center with which they are affiliated. For more information on NASA centers and facilities, see https://science.nasa.gov/ about-us/nasa-centers/.

Figure 2. Map of NASA Centers and Major Facilities.

[9] See, for example, House Committee on Science, Space, and Technology, Subcommittee on Space and Aeronautics, Enabling Mission Success from the Ground Up: Addressing NASA's Urgent Infrastructure Needs, hearing held July 29, 2021, https://www.govinfo.gov/content/pkg/CHRG-117hhrg45205/pdf/CHRG-117hhrg45205.pdf.

[10] FY2024 congressional budget justification, p. SSMS-35, https://www.nasa.gov/wp-content/uploads/2023/03/nasa-fy- 2024-cj-v3.pdf.

Personnel

Between the centers, headquarters, and other agency facilities, NASA has about 17,000 civil service employees.[11] The aging of the NASA workforce, retirement trends, and the alignment of skills with needs are issues of frequent congressional interest. Statistical data on employee age, retirement eligibility, job category, and other demographic factors are available on the Workforce Information Cubes for NASA (WICN) website.[12]

The number of civil service employees significantly understates the total NASA workforce. On-site and near-site contractor employees outnumber civil servants by about 2 to 1.[13] Nearly all staff at JPL, for example, are employees of Caltech, not NASA.

Budget and Appropriations

Figure 3 shows NASA's total budget since its establishment. The large peak in the 1960s funded the Apollo program. The smaller spike in FY1987 funded the construction of the space shuttle Endeavour to replace the shuttle Challenger after its loss in January 1986. Since that time, NASA funding has been relatively stable, after adjusting for inflation, with a gradual increasing trend over the past decade.

Proposed funding for NASA for the next fiscal year is included annually in the President's budget, released in early February (it is sometimes delayed).[14] NASA provides more detailed information, including program descriptions, status updates, and other data, in an annual congressional budget justification released with or shortly after the President's budget.[15] Congress acts on the Administration's request through the annual appropriations legislation developed by the House and Senate Appropriations Subcommittees on Commerce, Justice, Science, and Related Agencies (CJS). Bill language

---

[11] Based on full-time equivalents (FTEs). For a breakdown by center and appropriations account, see NASA's FY2024 congressional budget justification, p. SD-7, https://www.nasa.gov/wp-content/uploads/2023/03/nasa-fy-2024-cj-v3.pdf.

[12] See NASA, Workforce Information Cubes for NASA (WICN), https://wicn.nssc.nasa.gov/wicn_cubes.html.

[13] In FY2022, on a work-year equivalent (WYE) basis, NASA had 35,473 on-site or near-site contractor personnel performing recurring work for non-prime contractors. (NASA Office of Legislative Affairs, email to CRS, December 13, 2023).

[14] See Office of Management and Budget, "Budget," https://www.whitehouse.gov/omb/budget/.

[15] See NASA, "Budgets and Reports," https://www.nasa.gov/budgets-plans-and-reports/.

specifies appropriations for each NASA account and sometimes provides additional statutory direction, while language in accompanying committee reports and conference reports or explanatory statements typically provides extensive guidance about individual programs and projects. The CRS Appropriations Status Table tracks the progress of CJS appropriations legislation.[16] See also CRS Report R43419, *NASA Appropriations and Authorizations: A Fact Sheet*, by Daniel Morgan, which includes tables of NASA funding data for the year under current budget consideration as well as a few years of historical data.

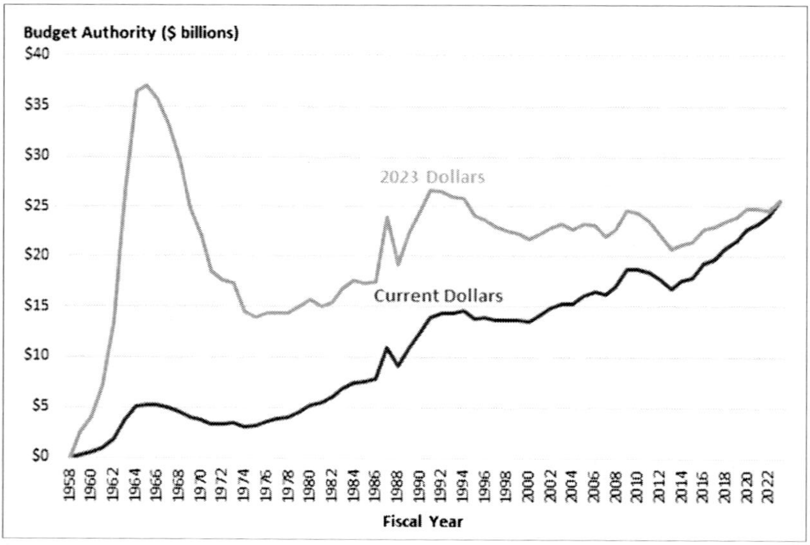

Source: Compiled by CRS. FY1958-FY2008 from National Aeronautics and Space Administration, Aeronautics and Space Report of the President: Fiscal Year 2008 Activities, Table D-1A, http://history.nasa.gov/presrep2008.pdf. FY2009-FY2022 from NASA congressional budget justifications, FY2011-FY2024, adjusted for supplemental appropriations, rescissions, and sequestration not shown in the justifications. FY2023 from P.L. 117-328 and explanatory statement, Congressional Record, December 20, 2022, pp. S7945-S7950. Current dollars deflated to FY2023 dollars using GDP (chained) price index from President's budget for FY2023, Historical Table 10.1, https://www.whitehouse.gov/omb/historical-tables/.

Notes: Transition quarter between FY1976 and FY1977 not shown.

Figure 3. Total NASA Budget, FY1958-FY2023.

[16] See CRS, Appropriations Status Table, https://www.crs.gov/AppropriationsStatusTable/.

Figure 4 shows NASA's FY2023 appropriations by account. The accounts for Science, Exploration, Space Operations, Space Technology, Aeronautics, STEM Engagement, and Inspector General fund the corresponding mission directorates and offices. Construction and Environmental Compliance and Remediation funds construction and environmental activities at NASA facilities. Safety, Security, and Mission Services (SSMS) funds a wide range of other supporting services, including operations and maintenance at NASA facilities. The SSMS account is currently the only NASA account that includes funding designated by Congress for community projects (sometimes known as earmarks).[17]

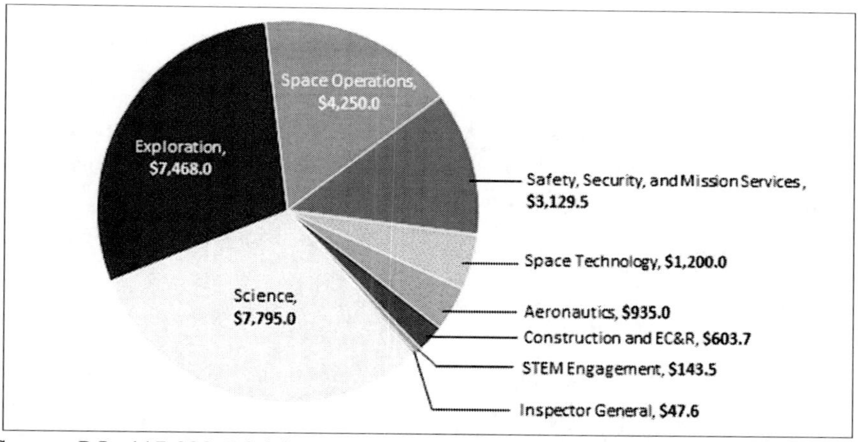

Source: P.L. 117-328, Divisions B and N, and explanatory statement, Congressional Record, December 20, 2022, pp. S7945-S7950.

Notes: STEM = science, technology, engineering, and mathematics; EC&R = environmental compliance and remediation.

Figure 4. NASA Appropriations by Account, FY2023. (Budget authority in millions)

In both the President's budget request and congressional appropriations legislation, funds are allocated by program and topic, not by NASA center or facility, even for projects that are managed largely or entirely at a single center. Each annual congressional budget justification, however, includes a table showing the anticipated distribution of funds from each appropriations account to each center.[18]

[17] For more information, see CRS Report R46722, *Community Project Funding: House Rules and Committee Protocols*, by Megan S. Lynch.

[18] See, for example, NASA's FY2024 congressional budget justification, p. SD-2, https://www.nasa.gov/wp-content/ uploads/2023/03/nasa-fy-2024-cj-v3.pdf.

Major Programs

NASA programs fall into five main categories: science, human spaceflight, crosscutting space technology development, aeronautics, and education. The two largest are science—conducted mostly with satellites and other uncrewed spacecraft—and human spaceflight. A separate program develops crosscutting space technologies to enable future missions. A program of aeronautics research seeks to advance flight within the atmosphere. An education program seeks to capitalize on NASA's scientific and technological capabilities to advance STEM education.

Science

NASA's Earth and space science activities are managed by the Science Mission Directorate (SMD) and funded though the Science appropriations account. They are organized into five programs: Earth Science, Planetary Science, Astrophysics, Heliophysics, and Biological and Physical Sciences. Each of these programs operates a fleet of satellites and other spacecraft, as shown in *Figure 5*.

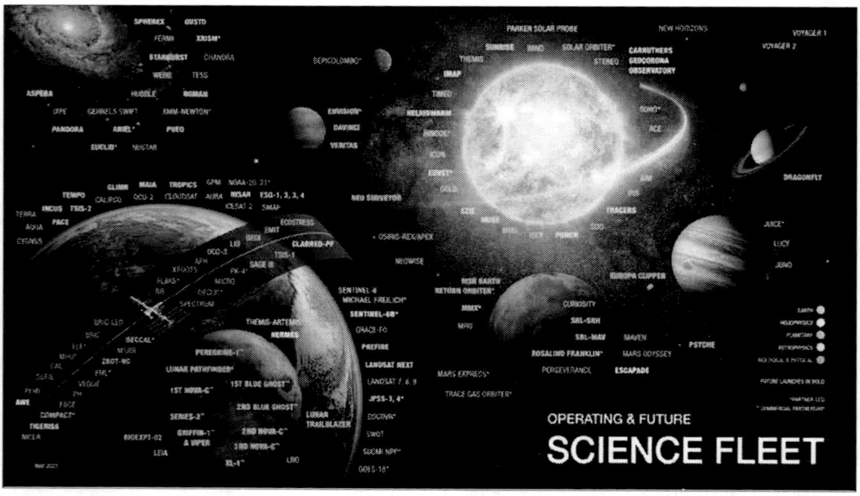

Source: NASA, Goddard Space Flight Center, https://svs.gsfc.nasa.gov/31162.

Figure 5. NASA Operational and Planned Science Missions (as of May 2023).

Earth Science

Earth Science conducts research on Earth's atmosphere, oceans, and land, as well as the biological and other processes that they undergo, primarily using observations from satellites in Earth orbit. The program also includes some aircraft-based projects. Climate science is a key motivator for this program, but the same satellites and sensors typically also provide data for research on a wide range of other topics, such as floods, droughts, wildfires, agriculture, and urban development.

The main focus of this program, like the other Science programs, is basic scientific research. The same imagery and other data, however, can sometimes also be useful for more applied purposes. Within Earth Science, the Applied Sciences subprogram facilitates the use of NASA satellite data, collected for research purposes, for use in applications such as response to natural disasters, drought monitoring, and agriculture.[19]

In partnership with the U.S. Geological Survey (USGS), an agency of the Department of the Interior, the Earth Science program develops and launches the Earth-imaging satellites of the Landsat program. After launch, Landsat satellites are operated by USGS. For more information, see CRS Report R46560, *Landsat 9 and the Future of the Sustainable Land Imaging Program*, by Anna E. Normand.

Planetary Science

Planetary Science conducts research on the Solar System's other planets, as well their moons, Earth's own Moon, asteroids, and comets. This work is conducted using robotic spacecraft that orbit or fly past those bodies or land on them for closer inspection. A Planetary Science mission of recent congressional interest is Mars Sample Return, which is intended to retrieve geological and atmospheric samples from Mars and return them to Earth for scientific study. An independent review board for the Mars Sample Return mission concluded in September 2023 that "there is currently no credible, congruent technical, nor properly margined schedule, cost, and technical baseline that can be accomplished with the likely available funding."[20] The National Academies of Sciences, Engineering, and Medicine has recommended that the Mars Sample Return mission "is of fundamental strategic importance to NASA, U.S. leadership in planetary science, and

---

[19] See NASA, Earth Science, Applied Sciences, https://appliedsciences.nasa.gov/.

[20] NASA, *Mars Sample Return (MSR): Independent Review Board-2 Final Report*, September 1, 2023, p.17, https://www.nasa.gov/wp-content/uploads/2023/09/mars-sample-return-independent-review-board-report.pdf.

international cooperation and should be completed as rapidly as possible. However, its cost should not be allowed to undermine the long-term programmatic balance of the planetary portfolio."[21]

Astrophysics

Astrophysics studies objects beyond the Solar System, such as galaxies, planets around other stars (known as exoplanets), and the history and evolution of the universe as a whole. This program includes familiar projects such as the Hubble Space Telescope, in Earth orbit, and the James Webb Space Telescope, in a distant orbit beyond the Moon. The James Webb Space Telescope encountered a series of high-profile budget and schedule challenges during development,[22] but was launched in December 2021 and is operational. Another Astrophysics mission, the Nancy Grace Roman Space Telescope (formerly the Wide-Field Infrared Survey Telescope, WFIRST), is scheduled for launch in 2027. An Inspector General report in 2021 found that its budget and schedule had been "significantly" affected (a $400 million increase and a six-month delay) by the COVID-19 pandemic.[23] In appropriations report language, Congress has expressed "the expectation that NASA will use a $3,500,000,000 development cost cap in execution of the mission."[24]

Heliophysics

The Heliophysics program studies the Sun itself as well as its impact on space weather and other phenomena elsewhere in the Solar System. It uses satellites in orbit around the Earth and the Sun as well as other more distant locations. Research on space weather through the Heliophysics program is part of a broader emphasis on space weather across multiple agencies in recent years, including an executive order issued in 2016, an interagency strategy and action plan issued in 2019, enactment of the Promoting Research and Observations of Space Weather to Improve the Forecasting of Tomorrow (PROSWIFT) Act

---

[21] National Academies of Sciences, Engineering, and Medicine, *Origins, Worlds, and Life: A Decadal Strategy for Planetary Science and Astrobiology 2023-2032* (2023), p. 3, https://nap.nationalacademies.org/catalog/26522/origins-worlds-and-life-a-decadal-strategy-for-planetary-science.

[22] See CRS In Focus IF10940, *The James Webb Space Telescope*, by Daniel Morgan.

[23] NASA, Office of Inspector General, *COVID-19 Impacts on NASA's Major Programs and Projects*, IG-21-016, p. 18, https://oig.nasa.gov/docs/IG-21-016.pdf.

[24] Explanatory statement accompanying the Consolidated Appropriations Act, 2023 (P.L. 117-328), *Congressional Record*, December 20, 2022, p. S7946.

(P.L. 116-181) in 2020, and an interagency research and operations framework issued in 2022.[25]

Biological and Physical Sciences

The smallest SMD program, Biological and Physical Sciences, uses the unique characteristics of space, such as microgravity, to conduct fundamental research in the biological and physical sciences. This research is done mostly using experiments conducted on or attached to the ISS. Before FY2021, the program was managed and funded as an element of the ISS program rather than as a separate activity within the Science Mission Directorate.[26]

Research and Analysis

Each of the five Science programs has its own Research and Analysis (R&A) program, devoted to supporting research based on data from existing spacecraft. The satellites and other spacecraft used in these programs are multiyear, multimillion-dollar (or more) missions. The cost of such missions is not uniform over time. Typically, costs are modest as a mission is being planned and designed, then ramp up rapidly for the manufacturing and launch of the spacecraft, then drop again once the spacecraft is operational. Funding for long-term operations and scientific data analysis nevertheless looms large in ensuring the scientific value of NASA's spacecraft investments. The availability of resources for R&A is a perennial focus of congressional oversight.

National Academies of Sciences, Engineering, and Medicine Decadal Surveys

NASA's Science programs rely heavily on a regular sequence of advisory reports from the National Academies of Sciences, Engineering, and Medicine (NASEM). Conducted approximately every 10 years for each major program area, these reports are known as decadal surveys. They provide guidance about scientific priorities, program management, and other matters, and are generally seen as reflecting the consensus of the U.S. scientific community.

---

[25] For more information, see CRS Report R46049, Space Weather: An Overview of Policy and Select U.S. Government Roles and Responsibilities, by Eva Lipiec and Brian E. Humphreys.

[26] For more on the organizational history of this program, see Marcia Smith, "Congress Approves Move of Biological and Physical Sciences from HEOMD to SMD," SpacePolicy Online.com, June 12, 2020, https://spacepolicyonline.com/news/congress-approves-move-of-biological-and-physical-sciences-from-heomd-to-smd/.

As well as a source of input, they are often used by NASA to justify actions it proposes to Congress, or conversely by Congress to justify direction it gives to NASA. Since the NASA Authorization Act of 2008 (P.L. 110-422), NASA's continued use of NASEM decadal surveys has been mandated in statute.[27] As well as these regular decadal reports, NASEM frequently provides other advice to NASA on particular topics outside the decadal cycle, sometimes at NASA's request, sometimes based on congressional direction.[28]

The most recent decadal surveys for the five Science programs are as follows:

- *Earth Science:* Thriving on Our Changing Planet: A Decadal Strategy for Earth Observation from Space (2018)[29]
- *Planetary Science*: Origins, Worlds, and Life: A Decadal Strategy for Planetary Science and Astrobiology 2023-2032 (2023)[30]
- *Astrophysics:* Pathways to Discovery in Astronomy and Astrophysics for the 2020s (2023)[31]
- *Heliophysics:* Solar and Space Physics: A Science for a Technological Society (2013)[32]
- *Biological and Physical Sciences*: Thriving in Space: Ensuring the Future of Biological and Physical Sciences Research: A Decadal Survey for 2023-2032 (2023)[33]

---

[27] 51 U.S.C. §20305.

[28] For example, in 2020, the Promoting Research and Observations of Space Weather to Improve the Forecasting of Tomorrow (PROSWIFT) Act (P.L. 116-181) mandated a NASEM review of a strategy developed by NASA and other agencies for observation of space weather. (51 U.S.C. §60602(c)).

[29] National Academies of Sciences, Engineering, and Medicine (NASEM), *Thriving on Our Changing Planet: A Decadal Strategy for Earth Observation from Space*, 2018, https://nap.nationalacademies.org/catalog/24938/thriving-on-our-changing-planet-a-decadal-strategy-for-earth.

[30] NASEM, *Origins, Worlds, and Life: A Decadal Strategy for Planetary Science and Astrobiology 2023-2032*, 2023, https://nap.nationalacademies.org/catalog/26522/origins-worlds-and-life-a-decadal-strategy-for-planetary-science.

[31] NASEM, *Pathways to Discovery in Astronomy and Astrophysics for the 2020s*, 2023, https://nap.nationalacademies.org/catalog/26141/pathways-to-discovery-in-astronomy-and-astrophysics-for-the-2020s.

[32] NASEM, *Solar and Space Physics: A Science for a Technological Society*, 2013, https://nap.nationalacademies.org/ catalog/13060/solar-and-space-physics-a-science-for-a-technological-society. For the progress of the next decadal survey of heliophysics, see https://www.nationalacademies.org/our-work/decadal-survey-for-solar-and-space-physics- heliophysics-2024-2033.

[33] NASEM, *Thriving in Space: Ensuring the Future of Biological and Physical Sciences Research: A Decadal Survey for 2023-2032*, 2023, https://nap.nationalacademies.org/

Joint Agency Satellite Division

Also within the Science Mission Directorate is the Joint Agency Satellite Division,[34] which manages the development and launch of weather satellites and other space-based instruments for the National Oceanic and Atmospheric Administration (NOAA, an agency in the Department of Commerce). NOAA reimburses NASA for these activities, which therefore appear in NOAA's budget, not NASA's.[35] Once operational after launch, the satellites are operated by NOAA directly.

Human Spaceflight

NASA's human spaceflight programs are managed by the Exploration Systems Development Mission Directorate (ESDMD) and the Space Operations Mission Directorate (SOMD), which are funded though the Exploration and Space Operations appropriations accounts, respectively. SOMD focuses on operational activities, particularly those associated with the ISS, while ESDMD focuses on the development of future capabilities, particularly those associated with the Artemis program to return humans to the Moon.

International Space Station

The ISS orbits at an altitude of about 240 miles and is composed of crew living space, laboratories, remote manipulator systems, solar arrays to generate electricity, and other elements. As well as providing facilities for research and technology development, it serves as a test bed for future human exploration missions to more distant destinations. Rotating crews from the United States, Russia, the European Union, and other countries have occupied it continuously since November 2000.

In an effort to increase use of the ISS by other federal agencies and the private sector, the NASA Authorization Act of 2005 (P.L. 109-155, §507) designated the U.S. portion of the ISS as a national laboratory. The NASA Authorization Act of 2010 subsequently directed NASA to contract with a

---

catalog/26750/thriving-in-space-ensuring-the-future-of- biological-and-physical-sciences-research-a-decadal-survey-for-2023-2032.

[34] See NASA, Science Mission Directorate, Joint Agency Satellite Division, https://science.nasa.gov/about-us/smd- programs/joint-agency-satellite-division/.

[35] See CRS In Focus IF12406, National Oceanic and Atmospheric Administration (NOAA) FY2024 Budget Request and Appropriations, by Eva Lipiec.

nonprofit organization to manage the ISS national laboratory (P.L. 111-267, §504). In 2011, NASA selected the Center for the Advancement of Science in Space (CASIS) as the ISS national laboratory managing organization.[36]

The framework for international cooperation on the ISS is the Intergovernmental Agreement on Space Station Cooperation, signed in 1998 by representatives of the United States, Russia, Japan, Canada, and 11 European countries.[37] The agreement is implemented through memoranda of understanding between NASA and its counterpart agencies in the other countries. Russia's participation at the operational level appears mostly unaffected by international tensions arising from Russia's 2022 invasion of Ukraine, despite some concerning statements from Russian officials in the early stages of the war.[38]

NASA used to rely on the space shuttle to carry U.S. cargo and crews to and from the ISS. The shuttle fleet was retired in 2011. Since 2012, two U.S. commercial providers—Space Exploration Technologies (SpaceX) and Northrop Grumman (formerly Orbital ATK)—have carried ISS cargo under NASA contracts. A third cargo provider, Sierra Nevada Corporation, also has a contract and is planning its first orbital launch and demonstration flight to the ISS in 2024. Between 2012 and 2019, ISS crews, including U.S. astronauts, were carried exclusively by Russian Soyuz spacecraft. NASA has crew transport contracts with two U.S. companies, SpaceX and Boeing.

The first operational crewed flight by SpaceX was in 2020. Boeing plans a test flight in 2024. To ensure cross-training for safety purposes, some U.S. astronauts still fly on Soyuz, and some Russian cosmonauts fly on NASA-contracted SpaceX flights, on a no-exchange-of-funds basis.

ISS operations were originally scheduled to continue only through FY2016. Statutory authority for continued U.S. operation of the ISS was extended through at least FY2020 by the NASA Authorization Act of 2010 (P.L. 111-267, §503(a)); through at least FY2024 by the U.S. Commercial Space Launch Competitiveness Act (P.L. 114-90, §114(b)); and through at least FY2030 by the NASA Authorization Act of 2022 (P.L. 117-167, §10815). At some point, even if operations are extended further, the ISS will reach the end of its useful life. To mitigate the risk of creating hazardous orbital

[36] See International Space Station National Laboratory, https://www.issnationallab.org/.

[37] See Intergovernmental Agreement on Space Station Cooperation, 1998, https://www.state.gov/wp-content/uploads/2019/02/12927-Multilateral-Space-Space-Station-1.29.1998.pdf.

[38] See, for example, Eric Berger, "The Western Space Community Should Put Dmitry Rogozin on 'Ignore,'" *Ars Technica,* May 2, 2022, https://arstechnica.com/science/2022/05/the-western-space-community-should-put-dmitry- rogozin-on-ignore/. At the time of the 2022 invasion, Rogozin was the head of Roscosmos, the Russian space agency.

debris, NASA plans to deorbit the ISS in a controlled manner.[39] It has issued a request for proposals for a Deorbit Vehicle, with contract awards anticipated in 2024.[40]

After the ISS, NASA expects to rely on commercially operated space stations to provide services and facilities in low Earth orbit (LEO). NASA's Commercial LEO Development program is supporting several companies to develop such capabilities. NASA has acknowledged that there may be a gap between the end of ISS operations and the first availability of commercial alternatives.[41]

Artemis

The core capabilities for the Artemis program are a crew capsule, known as Orion, and a heavy- lift rocket to launch Orion into space, known as the Space Launch System (SLS). Orion and the SLS have been in development since a congressional mandate in the NASA Authorization Act of 2010 (P.L. 111-267). An uncrewed test flight of Orion and SLS, known as Artemis I, was completed successfully in 2022. A first crewed test, known as Artemis II, is anticipated in 2024.

The crewed Artemis III mission, including the first human lunar landing since 1972, is anticipated in 2025.[42] Subsequent Artemis missions are planned.

Orion will not land directly on the Moon. Instead, a Human Landing System (HLS) is being developed to carry crews between lunar orbit and the lunar surface. NASA has contracted with two U.S. companies, SpaceX and Blue Origin, to provide HLS capabilities as a commercial service. NASA's decision to rely on commercial providers rather than NASA-owned spacecraft, and NASA's initial 2021 selection of SpaceX as the single provider, were both controversial in Congress. Since the award of a second contract to Blue Origin

---

[39] For more information, see NASA, "Frequently Asked Questions About the International Space Station Transition Plan," https://www.nasa.gov/faqs-the-international-space-station-transition-plan/.

[40] See NASA, "United States Deorbit Vehicle Contract," https://www.nasa.gov/johnson/jsc-procurement/usdv/.

[41] Jeff Foust, "NASA Acknowledges Possibility of Short-Term Post-ISS Gap," *Space News*, November 22, 2023, https://spacenews.com/nasa-acknowledges-possibility-of-short-term-post-iss-gap/.

[42] As of December 2023, NASA is reviewing this schedule. According to the Government Accountability Office, Artemis III is likely to be delayed to early 2027. (Government Accountability Office, *NASA Artemis Programs: Crewed Moon Landing Faces Multiple Challenges*, GAO-24-106256, November 30, 2023, https://www.gao.gov/ products/gao-24-106256).

in 2023, congressional attention has focused on the progress of HLS development and whether the system will be available in time for Artemis III.

For missions after Artemis III, NASA is developing a modular platform, known as Gateway, to be placed in a permanent orbit around the Moon. The first two Gateway modules—the Power and Propulsion Element (PPE) and the Habitation and Logistics Outpost (HALO, a pressurized habitat for astronauts)—are currently in development, with launch anticipated in 2025. Several of the planned subsequent modules will be contributed in-kind by international partners.[43] Gateway is intended to serve as a depot for storing supplies, a platform for science experiments, a location where subsystems launched separately can be assembled and integrated, and a rendezvous point where astronauts can transfer between Orion and the HLS and potentially, at some point in the future, depart for other, more distant destinations, such as Mars.

Other elements of Artemis include space suits, a surface habitat, a lunar rover, surface power systems, and so on. All these are in various stages of design and development.

The cost and schedule of the Artemis program overall, as well as of its individual elements, are a topic of continuing congressional scrutiny. At congressional direction,[44] the Government

Accountability Office has issued a number of reports reviewing the progress of SLS and of Artemis in general.[45]

For more information, see CRS In Focus IF11643, *Artemis: NASA's Program to Return Humans to the Moon*, by Daniel Morgan.

Space Technology

NASA's activities in crosscutting space technology development are managed by the Space Technology Mission Directorate (STMD) and funded though the Space Technology appropriations account. The program was established to focus on technologies that can enable future space missions, rather than those

---

[43] For more details, see NASA, "NASA's Gateway Program," https://www.nasa.gov/reference/nasas-gateway-program.

[44] For example, see H.Rept. 117-97, p. 134.

[45] For example, see Government Accountability Office, *Space Launch System: Cost Transparency Needed to Monitor Program Affordability*, GAO-23-105609, September 7, 2023, https://www.gao.gov/products/gao-23-105609; and Government Accountability Office, *NASA Artemis Programs: Crewed Moon Landing Faces Multiple Challenges*, GAO-24-106256, November 30, 2023, https://www.gao.gov/products/gao-24-106256.

needed for a specific mission currently under development. STMD has developed a strategic framework that organizes its investments and desired outcomes into categories such as advanced propulsion, in-situ resource utilization, and advanced avionics.[46] Technologies for space nuclear power and propulsion are a frequent focus of congressional interest.

STMD also manages the agency's Small Business Innovation Research (SBIR) and Small Business Technology Transfer (STTR) programs.[47] For more information on federal SBIR and STTR programs, see CRS Report R43695, *Small Business Research Programs: SBIR and STTR*, by Marcy E. Gallo.

Aeronautics

NASA's aeronautics activities are managed by the Aeronautics Research Mission Directorate (ARMD) and funded though the Aeronautics appropriations account. The program conducts research and development (R&D) on technologies for aviation, including aviation safety and environmental impact, as well as for air traffic management. Topics of frequent congressional interest include rotorcraft, hypersonic flight, electric propulsion and sustainable fuels, and the development of demonstrator aircraft, such as the X-59 Quiet Supersonic Technology (QueSST) experimental aircraft and the X-66A Sustainable Flight Demonstrator.[48]

The NASA Aeronautics Strategic Implementation Plan 2023 sets out ARMD's overall strategy, describes its role in six identified thrust areas, and discusses its needs for crosscutting research, workforce, and test capabilities.[49]

STEM Education

NASA's STEM education activities are mostly managed by the Office of STEM Engagement and funded though the STEM Engagement appropriations account. The Office has three main programs: the National Space Grant

---

[46] See NASA, TechPort, "Strategic Framework," https://techport.nasa.gov/framework.

[47] See NASA, Small Business Innovation Research (SBIR) and Small Business Technology Transfer (STTR) Program, https://sbir.nasa.gov/.

[48] For more information on supersonic passenger aircraft, including NASA's role, see CRS Report R45404, *Supersonic Passenger Flights*, coordinated by Rachel Y. Tang.

[49] See NASA, *NASA Aeronautics Strategic Implementation Plan 2023*, https://www.nasa.gov/wp-content/uploads/ 2021/04/sip-2023-final-508.pdf.

College and Fellowship Program (Space Grant), the Minority University Research and Education Project (MUREP), and the Established Program to Stimulate Competitive Research (EPSCoR). In addition, the Next Gen STEM program supports K-12 education and informal education at institutions such as museums and science centers.[50] These programs are authorized by Section 10851 of the CHIPS and Science Act (P.L. 117-167). Other STEM education activities are embedded in some of the agency's research programs, such as the Global Learning and Observations to Benefit the Environment (GLOBE) program, in Earth Science, and the Science Activation project, in Astrophysics.[51]

Space Grant

Space Grant supports education and student research through grants to consortia of colleges and universities.[52] There are 52 Space Grant consortia, one in each of the 50 states and one each in the District of Columbia and Puerto Rico. Consortia are selected competitively, usually every five years. In report language accompanying annual appropriations legislation, Congress often specifies a floor for the annual amount to be provided to each consortium. The Space Grant program was established in 1987 by the National Space Grant College and Fellowship Act (P.L. 100-147, Title II).[53]

MUREP

MUREP supports education and research at Minority Serving Institutions (MSIs), including Historically Black Colleges and Universities, Hispanic Serving Institutions, and other MSIs.[54] Funds are awarded competitively.

EPSCoR

NASA's EPSCoR program supports R&D at colleges and universities in states that have historically been less successful at competing for federal R&D

---

[50] See NASA, "Next Gen STEM for Educators," https://www.nasa.gov/learning-resources/for-educators/.

[51] See NASA, "GLOBE," https://www.nasa.gov/get-involved/globe/; and NASA, "Learn: Science Activation," https://science.nasa.gov/learn/.

[52] See NASA, National Space Grant College and Fellowship Project, https://www.nasa.gov/learning-resources/national-space-grant-college-and-fellowship-project/.

[53] For the act as amended, see 51 U.S.C. Chapter 403.

[54] See NASA, "MUREP," https://www.nasa.gov/learning-resources/minority-university-research-education-project/.

funding.[55] For more information on federal EPSCoR programs, see CRS Report R44689, *Established Program to Stimulate Competitive Research (EPSCoR): Background and Selected Issues*, by Laurie A. Harris.

Selected Other Topics

The remainder of this report consists of short sections addressing other focused topics of frequent congressional interest, including NASA's statutory authorities; where to find NASA policies and other Administration policies affecting NASA; space act agreements; where to find NASA funding data by state or district; NASA's regulatory role; the Artemis Accords; the Wolf Amendment prohibiting NASA collaboration with China; and major sources of external advice and analysis regarding NASA policies and programs.

Statutory Authorities

Most laws focused on NASA are codified in U.S. Code Title 51, which was created in 2010 to consolidate space-related statutes that had previously been in various other titles. The main exception is provisions derived from the NASA Authorization Act of 2010 (P.L. 111-267), which appear in 42 U.S.C. Chapter 159. Other relevant statutes are codified in a variety of titles. For example, statutes giving NASA special authorities regarding hiring and employment are codified along with related authorities for other agencies at 5 U.S.C. Chapter 98.

The foundation for most of Title 51 is the National Aeronautics and Space Act of 1958 (P.L. 85-568), the act that established NASA. Congress has enacted many amendments and additions since then, both in periodic broader-ranging NASA authorization acts and in focused legislation (e.g., the One Small Step to Protect Human Heritage in Space Act [P.L. 116-275] and the NASA Enhanced Use Leasing Extension Act of 2018 [P.L. 115-403]). Although NASA authorization acts are typically introduced and considered during most Congresses, they are not necessarily enacted on a regular annual or multiyear schedule, unlike some other authorization bills (such as the

---

[55] See NASA, "Established Program to Stimulate Competitive Research," https://www.nasa.gov/learning-resources/established-program-to-stimulate-competitive-research/.

defense authorization bill and the farm bill).[56] Two have been enacted in the past decade: the NASA Transition Authorization Act of 2017 (P.L. 115-10) and the NASA Authorization Act of 2022 (P.L. 117-167, Division B, Title VII).

Agency and Administration Policies

NASA has formal policies on a variety of topics, ranging from its own organization, management, and personnel to its procurement and the formulation and management of its programs. Current agency-wide policies are compiled and searchable at the NASA Online Directives Information System (NODIS).[57] They include high-level NASA Policy Directives (NPDs), more detailed NASA Procedural Requirements (NPRs), and guidebooks known as NASA Advisory Implementing Instructions (NAIIs), along with other less common categories.

Policies for NASA are also sometimes established at the Administration level through documents such as the National Space Policy, presidential Space Policy Directives, executive orders, and more focused national policies such as the Presidential Memorandum on Launch of Spacecraft Containing Space Nuclear Systems.[58] Such policies may be coordinated via the National Space Council, which consists of the NASA Administrator, the heads of other agencies with space activities, and other senior Administration officials, chaired by the Vice President.[59] They may also be issued by the Office of Science and Technology Policy (OSTP) in the Executive Office of the President, or the National Science and Technology Council (NSTC).[60]

---

[56] For a historical perspective, see the section on NASA in CRS Report R43862, *Changes in the Purposes and Frequency of Authorizations of Appropriations*, by Jessica Tollestrup.

[57] See NASA, NODIS Library, https://nodis3.gsfc.nasa.gov/main_lib.cfm.

[58] For a compilation of such policies, see NOAA, Office of Space Commerce, "Space Policies," https://www.space.commerce.gov/policy/.

[59] See The White House, National Space Council, https://www.whitehouse.gov/spacecouncil/. For the council's statutory authorities, see Section 501 of the NASA Authorization Act, Fiscal Year 1989 (P.L. 100-685) and other provisions set out as notes under 51 U.S.C. §20111. See also Executive Order 14056, *The National Space Council*, December 1, 2021, https://www.federalregister.gov/documents/2021/12/03/2021-26459/the-national-space-council.

[60] See, for example, NSTC, *National Preparedness Strategy for Near-Earth Object Hazards and Planetary Defense*, April 2023, https://www.whitehouse.gov/wp-content/uploads/2023/04/2023-NSTC-National-Preparedness-Strategy-and-Action-Plan-for-Near-Earth-Object-Hazards-and-Planetary-Defense.pdf. For more information on OSTP and NSTC and

Administration-level policies such as these are often high-level strategies or implementation plans with a focus on interagency coordination.

Space Act Agreements

At NASA's establishment, the National Aeronautics and Space Act of 1958 (P.L. 85-568), commonly known as the Space Act, authorized the new agency to "enter into and perform such contracts, leases, cooperative agreements, or other transactions as may be necessary in the conduct of its work and on such terms as it may deem appropriate."[61] The phrase "other transactions" allows NASA to craft agreements that are functionally contracts but are not subject to certain federal contracting requirements, such as the Federal Acquisition Regulation (FAR).

Such agreements are known as space act agreements (SAAs). Guidelines for the use of SAAs are contained in NASA's *Space Act Agreements Guide*.[62] A number of other agencies have subsequently received "other transaction" authority (OTA), usually with more detailed statutory guidance. The intent is typically to give agencies and companies more flexibility to negotiate the terms of an agreement, often when a company is paying all or part of the cost of its share of the agreed work.[63]

NASA Funding by State and District

The NASA Procurement Data View (NPDV) website allows any user to search for NASA contracts and other awards by contractor name or geographically by state and district.[64] Searches can be filtered by the type of award recipient (e.g., small business, educational institution, or nonprofit organization) and results can be presented either as dollar totals for the selected fiscal year or as detailed information on individual awards.

their relationship, see CRS Report R47410, *The Office of Science and Technology Policy (OSTP): Overview and Issues for Congress*, by Emily G. Blevins.

[61] 51 U.S.C. §20113(e).

[62] NASA, *Space Act Agreements Guide,* NASA Advisory Implementing Instruction NAII 1050-1D, https://nodis3.gsfc.nasa.gov/OPD_docs/NAII_1050_1D_.pdf.

[63] For more information on OTs, see CRS Report R45521, *Department of Defense Use of Other Transaction Authority: Background, Analysis, and Issues for Congress*, by Heidi M. Peters.

[64] See NASA, NASA Procurement Data View (NPDV), https://prod.nais.nasa.gov/cgibin/npdv/npdv.cgi.

Regulatory Role

NASA is generally not a regulatory agency, but it does issue regulations, such as 14 C.F.R. Chapter V (largely governing NASA's own operations and its interactions with other entities); 2 C.F.R., Subtitle B, Chapter XVIII (regarding requirements for NASA grants and other awards); and 48 C.F.R. Chapter 18 (the NASA supplement to the FAR).

The NASA Administrator serves on the FAR Council,[65] which was established by the Office of Federal Procurement Policy Act Amendments of 1988 (P.L. 100-679) to assist in the direction and coordination of procurement policy and procurement regulatory activities across the federal government. Under that authority, changes to the government-wide FAR (48 C.F.R. Chapter 1) are issued jointly by NASA, the Department of Defense, and the General Services Administration. In practice, though, FAR changes are initiated and developed via the Civilian Agency Acquisition Council (CAAC) and the Defense Acquisition Regulations (DAR) Council, for review by the FAR Council.[66] The CAAC and the DAR Council both have a broader base of agency participation than the FAR Council itself.

Note that NASA does *not* regulate commercial space activities, which generally fall under the regulatory jurisdiction of the Federal Aviation Administration, the Department of Commerce, and the Federal Communications Commission. For more information, see CRS Report R45416, *Commercial Space: Federal Regulation, Oversight, and Utilization*, by Daniel Morgan.

Artemis Accords

The Artemis Accords are agreements between NASA and its counterpart agencies in other countries, establishing "principles for cooperation in the civil exploration and use of the Moon, Mars, comets, and asteroids."[67] NASA and seven other countries signed the Accords in October 2020. More than 20 additional countries have signed since then. Negotiations with the partner

---

[65] See Acquisition.gov, Federal Acquisition Regulatory Council, https://www.acquisition.gov/far-council.

[66] See 48 C.F.R. §1.201-1.

[67] See NASA, "Principles for a Safe, Peaceful, and Prosperous Future," https://www.nasa.gov/artemis-accords/.

countries are co-led by NASA and the Department of State.[68] The Accords articulate principles, mostly based on the Outer Space Treaty of 1967,[69] rather than operational details.[70] NASA considers participation in the Accords a prerequisite for collaboration with NASA on the Artemis lunar exploration initiative, although the scope of the Accords is not limited to that. Note that the Artemis Accords are nonbinding agreements between NASA and other national space agencies; they do not have the status of a treaty.

Wolf Amendment (Prohibiting Cooperation with China)

The Wolf Amendment—named after its original author, former Representative Frank Wolf—is a provision in annual appropriations acts that prohibits most bilateral NASA cooperation with China. It was first enacted in 2011 as Section 539 of the Commerce, Justice, Science, and Related Agencies Appropriations Act, 2012 (P.L. 112-55, Division B). It has been repeated each year since then, with some variations from year to year. The FY2023 version prohibits bilateral NASA cooperation with China and the hosting of official Chinese visitors at NASA facilities, unless NASA certifies to Congress that an activity (1) poses no risk of technology or information transfer with national security or economic security implications and (2) will not involve knowing interactions with officials directly involved in human rights violations.[71]

The NASA Office of Inspector General issues an annual report on NASA's cooperation with China, listing any new or ongoing bilateral activities and confirming whether NASA made the required certifications to Congress for those activities.[72]

---

[68] See Department of State, "Artemis Accords," https://www.state.gov/artemis-accords/.

[69] See United Nations, Office for Outer Space Affairs, *Treaty on Principles Governing the Activities of States in the Exploration and Use of Outer Space, including the Moon and Other Celestial Bodies*, https://www.unoosa.org/oosa/en/ ourwork/spacelaw/treaties/ introouterspacetreaty.html.

[70] Some of the signatory countries also have more detailed operational agreements with NASA. See, for example, *Framework Agreement Between the National Aeronautics and Space Administration of the United States of America and the Israel Space Agency for Cooperation in Aeronautics and the Exploration and Use of Airspace and Outer Space for Peaceful Purposes* (2015), https://www.state.gov/wp-content/uploads/2019/02/15-1013-Israel-Space-Coop- Peaceful-Uses.pdf.

[71] See Section 526 of the Commerce, Justice, Science, and Related Agencies Appropriations Act, 2023 (P.L. 117-328, Division B).

[72] The 2023 report is NASA Office of Inspector General, *NASA's Compliance with Federal Export Control Laws*, IG- 23-009, February 6, 2023, https://oig.nasa.gov/docs/IG-23-009.pdf. For the Wolf Amendment, see pp. 1-2.

Major Sources of Assessment/Advice

As already mentioned, NASA makes extensive use of the National Academies of Sciences, Engineering, and Medicine for decadal surveys and other advisory studies (see "National Academies of Sciences, Engineering, and Medicine Decadal Surveys" above). These studies are mostly prepared under the auspices of NASEM's Space Studies Board and Aeronautics and Space Engineering Board.[73] It also receives external advice from the NASA Advisory Council (NAC) and the Aerospace Safety Advisory Panel (ASAP), both of which periodically issue recommendations and other reports and provide advice directly during regular open meetings.[74] The NAC and ASAP both operate under the requirements of the Federal Advisory Committee Act (FACA, 5 U.S.C. Chapter 10).[75] The NASA Office of Inspector General frequently reports on its audits and investigations, with recommendations for corrective actions.[76] The Government Accountability Office frequently publishes reports and testimonies on NASA, including a congressionally mandated annual report assessing major NASA projects,[77] other reports on individual NASA programs or topics,[78] and a searchable database of open recommendations.[79]

---

[73] See National Academies of Sciences, Engineering, and Medicine, Space Studies Board, https://www.nationalacademies.org/ssb/space-studies-board; and National Academies of Sciences, Engineering, and Medicine, Aeronautics and Space Engineering Board, https://www.nationalacademies.org/aseb/aeronautics-and-space- engineering-board.

[74] Reports, meeting minutes, and other materials are available online at NASA Advisory Council, https://www.nasa.gov/nac/, and NASA Aerospace Safety Advisory Panel, https://oiir.hq.nasa.gov/asap/.

[75] For more information, see CRS In Focus IF12102, *Federal Advisory Committee Act (FACA): Committee Establishment and Termination*, by Meghan M. Stuessy; CRS In Focus IF12512, *Federal Advisory Committee Act (FACA): Membership*, by Meghan M. Stuessy and Jacob R. Straus; and CRS In Focus IF12252, *Federal Advisory Committee Act (FACA): Meeting Requirements*, by Meghan M. Stuessy.

[76] See NASA Office of Inspector General, https://oig.nasa.gov/.

[77] The 2023 report is Government Accountability Office, *NASA: Assessments of Major Projects*, GAO-23-106021, May 31, 2023, https://www.gao.gov/products/gao-23-106021. This annual report series was mandated in the explanatory statement accompanying the Omnibus Appropriations Act, 2009 (P.L. 111-8). See *Congressional Record*, February 23, 2009, p. H1825. GAO defines major projects as projects or programs with a lifecycle cost exceeding $250 million.

[78] See GAO, Reports and Testimonies, https://www.gao.gov/reports-testimonies.

[79] See GAO, Reports and Testimonies, Recommendations Database, https://www.gao.gov/reports-testimonies/recommendations-database.

Chapter 2

International Collaboration and Competition in Space: Oversight of Nasa's Roles and Programs*

Committee on Commerce, Science, and Transportation

Thursday, October 21, 2021

U.S. Senate, Subcommittee on Space and Science,
Committee on Commerce, Science, and Transportation,

Washington, DC.

The Committee met, pursuant to notice, at 10:06a.m., in room SR–253, Russell Senate Office Building, Hon. John Hickenlooper, Chairman of the Subcommittee, presiding.

Present: Senators Hickenlooper [presiding], Cantwell, Blumenthal, Luja´n, Lummis, Wicker, Cruz, and Moran.

* This is an edited, reformatted and augmented version of Hearing Before the Subcommittee on Space and Science of the Committee on Commerce, Science, and Transportation United States Senate, Publication No. S. HRG. 117–816, dated October 21, 2021.

In: The National Aeronautics and Space Administration's Role …
Editor: Brent M. Jenkins
ISBN: 979-8-89530-539-3

Opening Statement of Hon. John Hickenlooper, U.S. Senator From Colorado

Senator HICKENLOOPER. The Subcommittee on Space and Science will come to order. Today marks this subcommittee's second hearing examining this, today, U.S. leadership in space. Timely and critical discussion—really timely and critical discussion, following recent launch of advanced Chinese aircraft.

We all recognize that NASA is a key agency for a number of programs, but especially important to civil and commercial space sector, enabling us to manage our international relationships in space. Congress last authorized a NASA authorization bill in 2017. We are absolutely thrilled that President Biden has confided in Senator Nelson and designated him to lead NASA, as the administrator. Together, hopefully in close cooperation, we can accomplish national goals within this, what we call, the last frontier.

Dating back into the Cold War, space competition was primarily between Americans and Soviets. Today, 40 nations have a space agency of some sort. Competition also exists between commercial space companies from a number of different nations.

The hearing today will examine several critical issues facing domain. One, how do we ensure the Artemis program remains competitive, on budget, on schedule, so that lunar landings can become a future reality that we can schedule with confidence? Are we going to be able to have spaceships—or space suits developed in time? The Artemis program is crucial to maintaining, what I think of as, the international confidence in America's ability to leader—to be a leader in space.

The next question, is how do we ensure NASA's long-term continuity of purpose, make sure that it is supported with sufficient appropriations by Congress, and can transcend Presidential administrations, so that we can get to secure funding that, not just this country, but the world will recognize as continuity?

Another question we will address is, is the U.S. future in low Earth orbit after the International Space Station, the ISS? The annual costs, now, is approaching almost $4 billion annually, just in operations. There are another commercial outposts under development, but there are still questions about how this will all shake out.

We are also going to look at how Space Traffic Management, STM, impacts U.S. missions in LEO, low Earth orbit. How do we mitigate orbital debris to make sure that we have sustainability in our commercial space industry? This subcommittee is going to have to develop new legislation to—to actively remediate some of the debris that is already up there. Which brings the questions, again, of how does space exploration and research get accomplished? How do we bring nations together in common purpose? How do we strengthen science, accelerate innovation and discovery?

Senate passed NASA reauthorization in the bipartisan U.S. Innovation and Competition Act, USICA—although I still prefer the Endless Frontier Act.

Senator WICKER. Here, here. [Laughter.]

Senator HICKENLOOPER. The Space and Science Committee will continue developing upcoming legislation to support NASA missions and make sure that we find that continuity of funding. I am delighted to welcome the panel. The Honorable Jim Bridenstine, former NASA Administrator under the Trump administration, former U.S. Representative from Oklahoma.

Dr. Mary Lynne Dittmar, from the Coalition for Deep Space Exploration. Someday I want to be associated with a coalition for deep space exploration. She served as Senior Policy Advisor for the ISS National Laboratory.

Mr. Mike Gold, of Redwire Space, former Associate Administrator for Space Policy and Partnerships at NASA.

And Dr. Patricia Sanders, Chair of the NASA Aerospace Safety Advisory Panel, former Executive Director of Missile Defense Agency.

I want to thank each of you for taking the time out of your busy schedules to join us here. I want to make sure that the conversation we have takes advantage of all of your diverse experiences and perspectives. I'm looking forward to the conversation.

Now, I will turn it over to Ranking Member Lummis for her opening comments.

Statement of Hon. Cynthia Lummis, U.S. Senator from Wyoming

Senator LUMMIS. Well, thank you, Mr. Chairman. It is so nice that we finally get to do this hearing. We were hoping to do it in Colorado Springs in August, and it would have been so nice to be in your beautiful state, then. But I am so delighted that we are getting to do this now. This is such a distinguished panel of witnesses and there is so much to learn for policymakers, and we rely so heavily on you to brief us, guide us, and support our efforts to sup port our space mission. So, thank you so much. I am so glad you got over your COVID and we are all back together again.

This group of panelists is so exceptional. So, thank you for your flexibility in working with us through the change and your willingness to be here today to speak with us. Your expertise is something we just welcome. So, very, very nice to have you here today and finally get to do this.

NASA has a long history of working with international partners and, of course, that is by design. The legislation that created NASA called for it to pursue cooperation with other nations and groups of nations. Our international partnerships have grown since NASA's creation and the success of the International Space Station speaks to the ways in which cooperation in space has strengthened our relationships on Earth. In fact, the first time I got to go to Russia—to Moscow, it was for the purpose of meeting with U.S. astronauts and Russian astronauts, who were there working together in Russia. Even in times of tension on Earth, American and Russian astronauts were able to work together at the International Space Station.

Now, five space agencies are part of the ISS project. The ISS will soon reach the end of its life span, but it cannot mark the end of our partnership. New opportunities and challenges await us, as we step in the era of the Artemis program and the Lunar gateway and eventually, human exploration of Mars.

While we embark on new adventures, we, along with every other spacefaring nation, must use them as opportunities to manage the challenge of space debris. We must work together to ensure we both remove orbital debris, and design equipment and policies to help cut back on the potential for orbital debris. I am encouraged by efforts in the private sector to step up and help with this problem, and I intend to work with my colleagues on solutions at the Congressional level to help tackle the issue.

There are other challenges, as well. Space is not only an area for collaboration. Competition in space looms large and we must work to avoid a cold war-style Space Race. Unfortunately, it seems some countries are less committed to this than others. Some nations do not seem interested in differentiating between military and civilian space operations. This week, we heard reports that over the summer, China launched a hypersonic glide vehicle that circled the Earth in low orbit and reentered the atmosphere. China's labeling this test of a nuclear capable weapon a "routine test of a space vehicle" underlines their unwillingness to separate military and civilian activities in space.

I suspect this trend will continue and it will become increasingly more difficult for the U.S. to view advancements made in space by China as anything other than a threat to our security. To that end, the United States must continue to pursue policies that make it the preferred partner for all other spacefaring nations and help to set norms that promote access, the sharing of scientific advancement, and neighborly attitudes toward other countries' assets in space.

So, I am very much looking forward to this hearing and learning from our distinguished panelists about these issues. I am sure we are going to learn so much today that I cannot wait to hear what you have to say. Thank you, Mr. Chairman. I yield back.

Senator HICKENLOOPER. Thank you, Ranking Member Lummis. Appreciate having Wyoming and Colorado run this committee. You have been a great partner and continue to be a great partner.

Now, I will recognize Ranking Member Wicker for his opening comments.

Statement of Hon. Roger Wicker, U.S. Senator from Mississippi

Senator WICKER. Thank you, Mr. Chairman. Senator Lummis, I came down with COVID the same time as Senator Hickenlooper.

[Laughter.]

Senator WICKER. I hope you are just as delighted that I got over it.

[Laughter.]

Senator WICKER. I guess it just goes without saying. As a matter of fact, Senator Hickenlooper, I found out how, finally, to get on the first few minutes of NBC Nightly News and that is to have an Independent and a Republican and a Democrat all come down with COVID on the same day.

[Laughter.]

Senator WICKER. This morning's hearing will address topics related to international cooperation and competition in outer space. Decades of U.S. cooperation in space, with other nations, has enhanced America's leadership in science and technology, and strengthened ties with international partners. We want to continue that.

NASA has been involved in thousands of cooperative activities with over 100 countries, including 20 years of sustained human presence aboard the International Space Station. Even during the height of the cold war, as Senator Lummis mentioned, space proved to be a rare area of cooperation between the U.S. and the USSR. Although space can bring nations together, the space domain has become increasingly competitive. Today, we are in a new space race with China. Beijing is seeking to secure both the lunar high ground, and low Earth orbit, LEO, to supersede the U.S. as the dominant space power. This year, China launched the first module of its space station and is eagerly soliciting international partners. Cooperative endeavors with other nations and the private sector will play a key role in helping America confront China's space challenge. NASA's Artemis deep space exploration program has already been critical in promoting U.S. leadership and cementing our alliances in space. Artemis will return astronauts to the surface of the moon and build—

Senator HICKENLOOPER. Here, here.

Senator WICKER. Cheers from the audience—and build a sustained presence there in preparation for future human exploration of Mars in this decade.

Rather than going it alone, the Artemis program envisions major contributions from international partners such as Canada, European nations, and Japan. In addition to relying on other nations, Artemis will also employ unique capabilities offered by commercial space companies for key aspects of its mission.

I look forward to hearing from our witnesses today on both the opportunities and the challenges NASA faces in competing against China, as well as carrying out the international commercial partnerships for Artemis, the International Space Station, and other programs.

NASA needs clear Congressional reauthorization and direction to keep Artemis on track and support other cooperative efforts. And I am delighted to report to you all that this is a priority of the Chair of the Full Committee, and I appreciate her efforts in this regard. The Committee's NASA Authorization Act, which passed as part of the Endless Frontier Act, includes many provisions to accomplish those goals. I am proud to have worked with Senator Cantwell and others on the NASA bill. I hope my colleagues in the House will work with us. Let us find a space vehicle to attach that NASA bill to and get it signed into law, at this pivotal time.

Thank you, Mr. Chairman.

Senator HICKENLOOPER. Thank you very much, Senator. We will have to find a space vehicle for the attachment.

Now, we can hear from out witnesses with their opening statements. I think we can start with the Honorable Jim Bridenstine.

Statement of Hon. Jim Bridenstine, Former NASA Administrator

Mr. BRIDENSTINE. Thank you, Chairman Hickenlooper, Ranking Member Lummis, Ranking Member Wicker, and Senator Moran. It is great to be back in Washington, DC, testifying before this committee.

I want to thank this committee for the strong, apolitical, bipartisan support that this committee has shown to NASA and its agenda. I think that is critically important for a lot of reasons. When you are doing projects that take multiple decades in nature, we have to have continuity of purpose. Continuity of purpose requires Congress to engage, be involved, and support the efforts of NASA in an apolitical, bipartisan way. And when that happens— this is the most important thing. When that happens, we are able to achieve international cooperation in a way that is often very difficult.

So, when you think about the history, you think about the vision for space exploration, the Constellation program, and the cancellation thereof. We think about the Space Exploration Initiative back in the 1990s and the cancellation thereof. When these things happen, it has—it has an effect on our international partners that is chilling. And they then choose, in many cases, not to partner with us and that is a problem.

But what I will say is that, if you look back at 2017, we passed the NASA Transition Authorization Act in bipartisan way, House and Senate. It was signed by the President in 2017. And, of course, the current NASA Administrator, then senator, came to that event, in a bipartisan way to say, look, NASA needs to have continuity of purpose. So, passing an authorization bill, in my view, is critically important. We have to have that to demonstrate to the world that we have resolve, as a Nation, to accomplish these very impressive things and to stay ahead of our competitors. So, that is, I think, probably the highest order of priority, getting that authorization passed.

Second thing I would like to say, and this is, I think, maybe even more important than the first, as when you think about the future, and that is the challenge of space debris. And Senator Hickenlooper, you nailed that right on the head. It is a big problem. And it is an even bigger problem when you think about the massive constellations that are being launched into low Earth orbit, even right now.

So, we think about how the FCC licenses these constellations. They look at a constellation—let us pretend it is 15,000 satellites large. And they say, there is a certain percentage of those satellites that are going to not be able to maneuver and not be able to reenter. In other words, they become derelict.

And they say—they apply a 1.5 percent chance for every satellite. So, if you launch 15,000 satellites that is 225 satellites that are derelict. And then, they say there is an aggregate collision risk for those 225 satellites, and they came up with a probability for a 15,000-satellite constellation that said there is a 1 in 44 chance of a collision in space. Here is the fundamental flaw with that. It only includes derelict satellites, which are 1.5 percent of the satellites. There is still a probability of collision for the other, you know, 98.5 percent of the thousands of constellations—satellites in that particular constellation. We do not know what that is. They have a maneuver capability—we do not know what that probability is.

But let us say there is a collision that is going to happen that is a 1 in 10,000 chance of a collision and if you maneuver, that probability goes down to 1 in a million. That is fantastic. We love that. We do not know that that is the case and so, we are licensing satellites right now, not knowing—it is a known unknown, and yet, we are still licensing the constellations, which I think is not a good idea.

But at the same time, here is what we do know. If it goes down to 1 in a million—which we do not know but let us pretend—and that eventuality happens 10 million times, that means there is going to be 10 collisions. So, if the probability goes down to 1 in a million and there are 10 million maneuvers, then there are 10 collisions. This is a—this is a mass—and by the way, I want to be clear. It goes beyond that. When we start talking about these constellations, we are only talking about satellites that we know that exist, and debris that we know that exists. For everything that we can track that is 10 centimeters or bigger, there are 10 to 100 things that we cannot track, that are equally lethal.

What I am saying is, that the challenge is much bigger than we know, but we know that we do not have any idea what that risk is. And yet, we are still launching these—and licensing these massive constellations without knowing what the risk really is. And I— I am telling you, it is much higher than what the FCC is currently predicting.

I will also tell you, on the 1.5 percent derelict satellites, it is— that is a— that is not right. It is going to be higher than 1.5 percent that cannot reenter.

Finally, I will say this. Just, I think it was, last night, somebody here can testify, Rwanda just filed for a constellation of 327,000 satellites into low Earth orbit. Recently, you had Spain file for a constellation that is 70,000 satellites. And of course, right now, we have got Starlink, and Kuiper and we have got OneWeb. We have got the European Union; they want to have their constellation. China, Russia—they all have these massive constellations for low latency, high throughput communications which, by the way, I support. But we do not want to destroy space in order to achieve it, which is what I am concerned about. The challenge is that there are—there are things that we absolutely know that we do not know, and we are still moving forward at a very rapid pace. And it is a very big concern, in my view. Orbital debris is number two. And I know that I am running out of time here quickly, but two other things that are important. I think it was mentioned by a number of senators, we have to have a replacement for the International Space Station. We love it. I know this committee passed an authorization bill out of the Senate that actually has it, no ki ding, at, you know, going to 2030. That is a good thing. I would also tell you that there is no guarantee it is going to last that long and China just launched a brand-new space station. That is another big problem. We cannot cede—we do not know the value of microgravity at this point, but what we do know is that, if we do not—if we lose the ISS and do not have a replacement, we are going to be in trouble.

Third thing, I think it is important to have two providers for the Human Landing System.

And with that, I will yield back. Thank you, Senator. [The prepared statement of Mr. Bridenstine follows:]

Prepared Statement of Hon. Jim Bridenstine, Former NASA Administrator

Chairman Hickenlooper, Ranking Member Lummis, and Distinguished Members of the Subcommittee,

It was my great pleasure to serve as the 13th Administrator of NASA. The men and women that serve our Nation at NASA are the best America has to offer. I would also like to commend the Biden Administration and specifically Administrator Bill Nelson for their commitment to NASA's continuity of purpose. NASA's projects are multi-decadal and even multi-generational in nature. History has testified that whimsical shifts based on administration changes result in billions of dollars wasted and limited progress as our competitors and adversaries advance resolutely. Political leaders who care about exploration, science, discovery, and American leadership must constantly work to keep NASA outside of partisan political wrangling. I worked to that end every day and Administrator Bill Nelson has been working toward the same objective.

There are three specific challenges facing NASA with which this committee can help:

1. Human exploration in low Earth orbit
2. Orbital debris mitigation
3. Redundancy in America's Moon lander

Low Earth Orbit

It is absolutely critical that Congress fund America's replacement for the International Space Station (ISS), which has had humans onboard for over 20 years. The Chinese Space Station is on orbit, demonstrating great progress, and attracting international partners. Humanity is only at the beginning of understanding the immense economic, technological, and medicinal value of microgravity and America is at risk of ceding these capabilities to our greatest competitor. Congress must not let this happen.

America's advantage is its capital markets that are funding commercial human habitation and the brilliant entrepreneurs that are putting that capital to work. While the future of human spaceflight in low Earth orbit (LEO) is commercial, the United States Government has an interest in ensuring the success and steering the direction of these capabilities. Congress needs to fund NASA's LEO commercializa tion efforts at $2 billion per year. This money should be used for development and demonstration of LEO habitation as well as establishing NASA and its partners as ongoing customers of commercial LEO habitation. If Congress does this, capital markets and entrepreneurs will respond in a way that establishes America as pre-eminent in LEO human spaceflight at a cost significantly less than the ISS.

Once LEO commercialization is funded, Congress should require NASA to establish a date for the transition to new space stations with a plan to ensure there is no gap in American human spaceflight in low Earth orbit.

Space Debris Mitigation

Satellite communications are advancing rapidly and will be transformative for humanity. New constellations in geostationary orbit (GEO) are breaking records for capacity and throughput, while drastically dropping the cost per megabit. Constellations being developed for low Earth orbit (LEO) could complement these systems by delivering low latency communications allowing for applications such as tele-robotics and video gaming. Global, hybrid networks (GEO and LEO) will soon deliver additional benefits to online education, telemedicine, mobile banking, and economic opportunity for people around the world waiting to connected.

However, the U.S. government and governments around the world are failing to properly manage collision risk. If not remedied, the consequence will be losing access to space entirely, devastating not only satellite communications, but also human spaceflight, national security, weather prediction, disaster relief, climate science, and so much more.

The challenge is that we are beginning an era of large, multi-satellite constellations when FCC rules on debris mitigation apply on a satellite-by-satellite basis. Requiring a failure probability of less than 1/1000 over 5 years makes sense for a single satellite. However, if 40,000 satellites are launched into the same orbital shell, each with a probability of failure of 1/1000, there will be 40 failures on average. In short, the current requirement for low probability of failure becomes the inevitability of numerous failures.

The problem becomes much worse when considering collisions. A study by the MITRE Corporation, commissioned by the National Science Foundation (NSF), suggests that it will not be feasible to operate some of the LEO systems being planned. That study predicts the effects of a LEO system that plans to include over 40,000 satellites at an altitude of about 600 kilometers.

As shown in the following figure from that study, dramatic increases in space collisions, and new space debris, are expected within just a few years. In the longer term "satellites are destroyed [by debris and collisions] faster than they are launched."[1]

The challenge increases drastically when considering that there are dozens of companies in the United States building constellations all desiring access to LEO. Additionally, China, Russia, Europe, and individual European countries have announced they will be deploying LEO constellations as well. The United States can and must take the lead on creating responsible rules that others in the international community can emulate to mitigate this growing concern.

---

[1] G. Long, The Impacts of Large Constellations of Satellites, JASON—The MITRE Corporation, JSR–20–2H, November 2020, (Updated: January 21, 2021), at 97, available at *https://www.nsf.gov/news/special_reports/jasonreportconstellations/JSR-20-2H_The_Impacts_of_Large_Constellations_of_Satellites_508.pdf.*

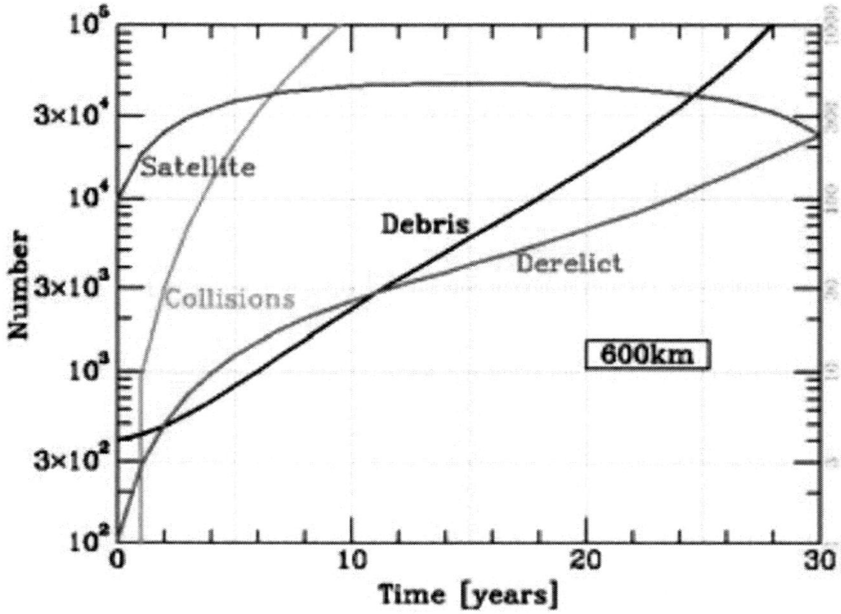

Space situational awareness and space traffic management (SSA/STM) are important long-term requirements. However, if we don't take action now to mitigate the debris problem, there will be no need for SSA/STM. Space will no longer be accessible. I fully support the Commerce Department being responsible for SSA/STM, but given the immediacy of the problem, we need to use our current framework for mitigation.

Solution:

First, Congress needs to quickly require the FCC to approve satellite constellations based on an aggregate collision risk metric, not one satellite at a time. The FCC proposed this rule in April 2020, but it was never implemented. It was the right policy then and it is even more the right policy today.

Second, Congress must act quickly to require the FCC to define the limits on the nature and number of satellites that can exist in LEO. The FCC has said there are limits, but there has been no action.

Third, once those limits are established, there must be an effort to work with our international partners to appropriately allocate access to the highest value orbital shells. If other nations are not included in this process, they will claim the same territory with disastrous consequences. America's allocation must then be fairly distributed to companies in a way that ensures competition in the marketplace.

Finally, Congress should have the FCC analyze and report on the effects these constellations will have on launch, the International Space Station, the environment, astronomers, and future space exploration.

Artemis: America's Moon Program

NASA determined that it had insufficient appropriations to select two human landing systems for the Artemis program. Congress should appropriate sufficient funds for two human landing systems.

Dissimilar redundancy has worked exceptionally well in the Commercial Resupply Program and the Commercial Crew Program. When one contractor has a setback, the other moves forward, greatly increasing NASA's access to space and the International Space Station. This should be the approach for NASA's Human Landing System and Congress needs to fund it.

Additionally, if the goal is to eventually buy access to the Moon as a commercial service, NASA needs two providers that will compete on cost, innovation and safety. Having two providers will also incentivize the companies to get customers that are not the U.S. Government, lowering the cost to the taxpayer. As a matter of principle, the only thing worse than a government monopoly is a private monopoly to which the government is beholden. I am not suggesting that any specific contractor would intentionally act nefariously. I am suggesting that the government has an obligation to provide the right incentives to benefit the taxpayer.

Thank you for hearing my testimony. I am confident that with your leadership, NASA will continue to do amazing things for our Nation.

Senator HICKENLOOPER. Thank you, Jim, appreciate that. Dr. Dittmar.

Statement of Dr. Mary Lynne Dittmar, Executive Vice President for Government Affairs, Axiom Space, Inc.

Dr. DITTMAR. Chair Hickenlooper, Ranking Member Lummis, and Senator Moran, Ranking Member Wicker, Members of the Sub-committee, thanks for the invitation to appear before you today. It is nice to be back in front of this committee with such a distinguished panel to discuss my thoughts on the topics of today's hearing.

I am the former President and CEO of the Coalition for Deep Space Exploration, but I am now, the Executive Vice President of Axiom Space, which is building the world's first commercial space station. So, I wish to note that while I will be discussing some of Axiom's plans as to illustrate my discussion, the positions that are presented here are my own and do not reflect the official opinion of my employer.

With that out of the way, my discussion will focus on competition—that aspect of today's hearing. It is often said that we are at an inflection point in space and while that is true, the phrase glosses over complexities, challenges, and opportunities, already pointed out, that confront us now and in the years to come.

What is less frequently discussed is a more dangerous reality— namely, that if the U.S. does not act decisively and strategically, it risks squandering the competitive moat that it has built in commercial space to national actors who continue to maintain laser focus on asserting their own national interests.

Axiom is the first, and so far, only company to develop a new space station destined for low Earth orbit without government funds for development, launch, or operations. Beginning by building modules that extend the functionality of the ISS, it will provide capabilities for a broad array of users—researchers, astronauts, government, international customers, and the business community. Critically, for the purposes of this hearing, it is also, in my opinion, best positioned to help challenge China's interests in LEO at the end of life of the International Space Station. And indeed, it was for that latter reason, more than any other, that I joined Axiom. It is tremendously important for the U.S. Government to recognize and respond, with urgency, to the situation our Nation and space industry is facing. Simply put, it is China's goal to establish a leading position in the economic and military use of outer space, as they have stated publicly, for many years.

China recognizes the immense strategic value of space and intends to seize the opportunity to develop economic value, by dominating a space-based economy. Space has become part of China's Silk Road economic belt, part of its plan for geopolitical and economic dominance.

In my written testimony, I refer to the U.S.-China Economic and Security Review Commission's 2019 report to Congress, which offers the comprehensive analysis of what Beijing calls its "Space Dream" and how it hopes to achieve it. China is playing the long game, make no mistake. Bolstering their space capabilities through military-civil fusion and through the exploitation of weaknesses in the U.S. space industrial base and our glacial pace of acquisition and procurement and of the stovepipe nature of our national agencies.

With regard to LEO, NASA, and by extension its stakeholders, are facing uncertainty. Plans to transition off the ISS at the end of its life require clearly stated objectives, realistic timelines with milestones, and a firm commitment that has not yet been adequately communicated. At present, NASA is describing its objectives partly in terms of continued access to LEO, envisioning hardware development under its new commercial LEO Destinations Program, using the same acquisition strategy that works so well for Commercial Orbital Transportation System, COTS, and eventually for commercial crew.

There are several differences between the circumstances that gave raise to COTS over a decade ago, however, and those in low Earth orbit now. COTS was able to leverage an existing market for launch services, that had developed and diversified over 50 years, to include an understanding of hardware development and a substantial customer portfolio in both government and commercial sectors. COTS was also set up to meet transportation needs that were communicated early in the program.

Today, conditions are markedly different. While now, as then, hardware development is relatively well understood, there is no diversified market in LEO to leverage. Not yet. NASA has yet to clearly define its needs for services after the ISS ends, nor does it plan to do so for some time.

And finally, we face a formidable competitor. At the very same time that NASA is embarking on this approach, American companies have admitted to losing customers to China as it employs strategies to undercut the U.S. commercial space sector through mechanisms such as state-backed financing that market driven companies in the U.S. cannot compete with. In my view, an important national objective for LEO should be to counter potential Chinese hegemony in low Earth orbit. Should Congress choose to authorize LEO programs with this goal in mind, then NASA's acquisition approach should reflect this. It is unclear that the same procurement approach that worked for COTS, in very different circumstances, will work now.

I want to spend just a few minutes talking about demand. Both industry and NASA believe that the market sector is ready to develop next generation orbital platforms. In Axiom's case, this confidence is shared by capital markets. In a few years, Axiom will see its modules attached to station, increasing capacity and capabilities for meaningful science and research. And I have got a quick visualization, if that is available to us to see, so you can see how that will go. Maybe not. OK, moving on. However, Axiom and other companies providing services in LEO, now or in the future, are dependent upon sufficient demand to close their business cases. If there is not sufficient demand, then the very companies the U.S. is depending on to assure U.S. presence and soft power in low orbit may fail.

In our system, governments do not create markets. They can only serve as customers. However, NASA can fund and conduct research that is too costly, risky, or difficult for industry to undertake and provide the results to industry for use in engaging with customers. If market development is to be a legitimate aim of space policy, much as it is for simple state and commerce, particularly if it is to underpin pursuit of a range of national, as well as business goals, we might also ask if NASA is the appropriate home for that policy. NASA is not an economic development agency, and we should not expect it to act as one. The space agency, that serves this country so well, has core confidence revolving around science, exploration, education, and technology.

At the same time, as a Nation, we must think and act deliberately, with regard to the intersection between markets and space policy. In my written testimony, I offer several suggestions and recommendations that the committee may consider, regarding how government may help.

It has been a pleasure and an honor to appear before you today and I look forward to answering any questions you may have.

[The prepared statement of Dr. Dittmar follows:]

Prepared Statement of Dr. Mary Lynne Dittmar, Executive Vice President, Axiom Space, Inc.

Chair Hickenlooper, Ranking Member Lummis, Chair Cantwell, Ranking Member Wicker, and members of the Subcommittee, thank you for the invitation to appear before you today with such an extraordinary and distinguished panel to discuss my thoughts on the topic of today's hearing: "International Collaboration and Competition in Space: Oversight of NASA's Role and Programs." My name is Dr. Mary Lynne Dittmar, and I am the former President and CEO of the Coalition for Deep Space Exploration—an industry trade group supporting NASA's programs of record in human exploration, science, and space commerce—and now, the Executive Vice President of Axiom Space, which is building the world's first commercial space station. The topic of this hearing is of intense interest to me, as it has guided most of my professional activities and service for more than 20 years.

I wish to note that while I will be discussing some of Axiom's plans to illustrate my discussion, the positions presented here are my own and do not reflect the official opinion of my employer.

Axiom is the first (and so far, the only) company to develop a new station destined for low Earth orbit (LEO) without government funds for development, launch, and operations. Beginning by building modules that extend the functionality of the ISS, it will provide capabilities to a broad array of users—researchers, astronauts, manufacturing firms, scientists, artists—and will offer those capabilities to meet existing and emerging needs across the U.S. government, international customers, and the business community. Critically, for the purposes of this hearing, it is also best positioned to help challenge China's interests in LEO at the end of life of the International Space Station. Indeed, it was for that latter reason, more than any other, that I joined Axiom.

It's often said we're at an "inflection point" in space, and while that is true, the phrase glosses over the complexities, challenges and opportunities confronting us now and in the years to come. Regarding LEO, the term has been used to describe the transition from an exclusively government owned-and-operated domain to one utilizing publicprivate partnership. What is less frequently discussed is a more dangerous reality—namely, that if the U.S. does not act decisively and strategically, it risks squandering the competitive moat the U.S. has built in commercial space to national actors who continue to maintain laser focus on asserting their own national interests. In this testimony I will discuss this situation in both LEO and in deep space, with an emphasis on human spaceflight and exploration.

NASA's Efforts to Build Alliances

The National Aeronautics and Space Act of 1958—the so-called "organic act" that created NASA—has as one of its nine objectives "Cooperation by the United States with other nations and groups of nations in work done pursuant to this Act and to the peaceful application of the results thereof."[2] In other words, cooperation with other nations is in the DNA of our Nation's space

[2] U.S. Congress (1958). National Space and Aeronautics Act of 1958, Pub. L. No. 85–568, as amended (29 July). Printed for NASA Use November 30, 2006. Accessed at: *https://history.nasa.gov/spaceact-legishistory.pdf.*

agency. This part of NASA's mission has been ratified by Congresses since 1958 in a series of Authorization Acts that build upon the first.

NASA has demonstrated its commitment to that cooperation in the years since— in space science, in human spaceflight, in technology development, and in education and outreach. The most visible example—but certainly not the only one—of multilateral cooperation in space is the International Space Station (ISS). For the past 20 years the ISS has been the focus of a strong international partnership that has attracted over 100 countries as participants. NASA's daily engagement with the partnership has weathered changing political climates both at home and abroad, demonstrating extraordinary stability, positive interdependence, and an ability to bridge the tensions and conflicts that inevitably surface in international relations.[3] This has been particularly true with Russia, which served to ensure the continuation of the ISS after the loss of the shuttle Columbia and in the long interval between 2011 and the end of the shuttle program until 2020 when the U.S. was again able to launch astronauts on American rockets from American soil. This partnership has been even more remarkable given the stresses that attended the Russian annexation of Crimea, and recent, sometimes contrary statements from Roscosmos regarding Russian intentions for future participation on the ISS.

Guiding this cooperation has been the Intergovernmental Agreement (IGA), a treaty-level multilateral agreement among the governments of Canada, governments of member states of the European Space Agency, the government of Japan, the government of the Russian Federation, and the government of the United States.[4] The 15 member nations who signed the IGA after years of negotiations have remained committed to its goals and principles and have created a strong foundation of multilateralism in human spaceflight. However, the IGA is specific to the ISS and does not extend to exploration of deep space. In addition, China was not included in the ISS partnership.

In 2010, Congress directed NASA to ask the National Academies to perform a study to review "the goals, core capabilities, and direction of human spaceflight." In 2012 NASA asked the National Research Council at the Academies to perform a wide-ranging study in accord with Congressional

---

[3] Dittmar, M. L. (2017). Statement before the Committee on Space, Science and Technology, U.S. House of Representatives, 22 March. ''The ISS After 2024: Options and Impacts.'' *https://docs.house.gov/meetings/SY/SY16/20170322/105737/HHRG-115SY16-Wstate-DittmarM-2017 0322.pdf.*

[4] The U.S. State Department (1998). Space Station Agreement between the United States of America and Other Governments (signed 29 January 1998). *https://www.state.gov/wp-content/ uploads/2019/02/12927-Multilateral-Space-Space-Station-1.29.1998.pdf.*

direction. The NRC in turn convened a Committee on Human Spaceflight, on which I served, to carry out the work. The resulting report, entitled "Pathways to Exploration: Rationales and Approaches for a U.S. Program of Human Space Exploration," was published in 2014.[5] Among many other recommendations, a central finding of the Committee was that NASA's deep space exploration program must be international in nature, and that continuing engagement is necessary to develop and maintain international partnerships. The report also recommended that China be engaged, arguing that exclusion of the Chinese space program could work against the interests of the United States, particularly given China's stated openness to working with other nations.[6]

In October of 2020, NASA announced the signing of the Artemis Accords by eight founding member countries including the United States. Announced earlier in May of that year, the Accords "reinforce and implement" the principles of the Outer Space Treaty, and are aimed at avoiding conflict, strengthening international rela tionships, and encouraging ongoing cooperation in a growing coalition dedicated to the peaceful uses of outer space. The Accords do not constitute a treaty, but rather a framework to encourage bilateral agreements developed under the umbrella of several principles that, it is hoped, will develop into international law through custom and practice. The Accords thus offer a significant step toward implementing the recommendation of the Pathways report that U.S. deep space exploration be international in nature, providing a framework of principles underpinning future exploration: Transparency, interoperability, emergency assistance, registration of space objects, release of scientific data, preservation of outer space heritage, extraction and utilization of space resources, deconfliction of space activities, and planning for the safe disposal of orbital debris.[7]

---

[5] National Research Council of the National Academies Human Spaceflight Committee (2014). Pathways to exploration: Rationales and approaches for a U.S. program of human space exploration. Washington, D.C., National Academies Press.

[6] Ibid, p. 26.

[7] National Aeronautics and Space Administration (2020). NASA, international partners advance cooperation with first signings of Artemis Accords. 13 October. *https://www.nasa.gov/press-release/nasa-international-partners-advance-cooperation-with-firstsignings-of-artemis-accords.*

Progress and Intentions of Other Nations in Space: Focus on China

It is worth noting that neither the People's Republic of China (PRC) nor Russia have announced an intention to sign the Artemis Accords, offering mixed responses to the initiative at various times.[8,9] However, the two countries have signed an intergovernmental agreement to proceed with plans for an International Lunar Research Station (ILRS). In June they announced opportunities for additional partnerships and said that they were already in negotiations with several nations.[10] China, in particular, shares the U.S. determination to assert leadership in cislunar space, recognizing its immense strategic value and seeing in it an opportunity to develop economic value by dominating a space-based economy.[11]

To fully assess China's intentions in space, it is important to understand how the PRC is advancing with extraordinary rapidity across a range of space technologies and capabilities. Notably, China is "playing the long game," focusing on the relationship between the defense sector and commercial space innovation and entrepreneurialism and leveraging the existing social commitment to hard work and national pride. Space has become part of China's "Silk Road Economic Belt," part of its plan for geopolitical and economic dominance. In short, the PRC is pursuing a global strategy to win the competition in space, building upon several broad-based approaches.

The first of these is via military-civil fusion (MCF). MCF is focused on building pathways between the traditional defense industrial base and the commercial sector, with the goal of increasing the flow of information, technology, and people between the two. This is not a new development in China—or for that matter, in the United States—but it has new impetus under

---

[8] TASS Russian News Agency (2020). Moon exploration projects could help Russia-US coopera- tion—Roscosmos. 16 May. *https://tass.com/science/1157255.*

[9] Ji, E., Cerny, M. B., Piliero, R. J. (2020). What does China think about the Artemis Accords? The Diplomat, 17 September. *https://thediplomat.com/2020/09/what-does-china-think-about-nasas-artemis-accords/.*

[10] Pultarova, T. (2021). Russia, China reveal moon base roadmap but no plans for astronaut missions yet. 17 June. *https://www.space.com/china-russia-international-lunar-research-station.*

[11] U.S.-China Economic and Security Review Commission Annual Report to Congress (2019). China's ambitions in space: Contesting the final frontier (Chapter 4). *https://www.uscc.gov/sites/default/files/2019-11/2019%20Annual%20Report%20to%20 Congress.pdf.*

the current government.[12] According to testimony provided in 2019 by the U.S.-China Economic and Security Review Commission, in 2013 there was virtually no commercial space sector in China. In 2014, the government announced that it would allow the flow of private capital and companies to enter the previously sequestered space sector. In 2016, the Central Politburo elevated MCF to a national strategy. In 2017, President Xi Jinping established a commission for Integrated Military and Civilian Development emphasizing technology development—particularly dual-use technology. Later in 2017, the People's Liberation Army declassified almost 3000 patents, releasing them into the public sector, signaling a new development in the interaction between the PLA and the commercial sector. Also in 2017, the "13th Five Year Plan Sci & Tech MilitaryCivil Fusion Development Special Plan" named human spaceflight as a MCF "mega project."[13] By 2018, the top seven state funds investing in MCF industries had over $56B in capital. Combined with capital from venture markets, the Chinese government is pouring billions into commercial space companies as compared with just a few years ago.

While the total investment in both government and commercial space is thought to still lag that of the United States, the results are clear to see. For example, China's commercial launch industry has made rapid progress since 2015, with several companies testing advanced rocket capabilities, including fly-back boosters, sub-orbital flights, and advanced engine designs, including a liquid oxygen-methane engine.[14] Notably, the PLA remains deeply engaged in guiding technical development, particularly those capabilities that it sees of benefit to it. Further, the personnel and relationships between the commercial space sector and the state defense sector are also deeply intertwined.[15]

A second means by which the PRC seeks to compete with the U.S. is taking shape through a long-term, multipronged strategy that exploits weaknesses in the U.S. space industrial base, in our glacial pace of acquisition and procurement, and of the stove piped nature of our national agencies. We

---

[12] Kania, A.B. and Laskai, L. (2021). Myths and realities of China's military-civil fusion strategy. Center for a New American Security. *https://s3.us-east-1.amazonaws.com/files.cnas.org/documents/Myths-and-Realities-of-China's-Military-Civil-FusionStrategy_FINAL-min.pdf? mtime=20210127133521&focal=none.*

[13] Laskai, L. (2019). Testimony before the U.S.-China Economic and Security Review Commission, 25 April. "China in Space: A Strategic Competition?" *https://www.uscc.gov/sites/default/files/Lorand%20Laskai%20USCC%2025%20April.pdf.*

[14] Curcio, B. and Lan, T. (2018). The rise of China's private space industry. Space News, 25 March. *https://spacenews.com/analysis-the-rise-of-chinas-private-space-industry.*

[15] Lee, T. A. and Singer, P. W. (2021). China's space program is more military than you might think. 16 July. Defense One. *https://www.defenseone.com/ideas/2021/07/chinas-space-program-more-military-you-might-think/183790/.*

have reached the point where a whole-ofgovernment approach to space, as intended to be implemented via the National Space Council among other mechanisms, is crucial to our continued competitiveness and leadership. This is particularly true regarding protecting the space industrial base and the emerging entrepreneurial sector, both of which are vulnerable to malfeasance and disruption—the latter clearly illustrated by impacts of the COVID–19 pandemic. In Q&A during a hearing[16] on competition with China in space before the U.S.-China Economic and Security Review Commission, Mike Gold pointed out the PRC thinks about markets strategically, while the U.S. does not. China has an industrial policy organized around state objectives. The U.S., with its focus on free markets, does not.

The 2019 Report to Congress of the U.S.-China Economic and Security Review Commission pointed out the need for a comprehensive, wholistic perspective regarding China's competitive interests in space.[17] A recent report of a workshop in early 2019 co-sponsored by the Air Force Research Laboratory (AFRL) and the Defense Innovation Unit (DIU) detailed a methodical, intentional approach that includes deliberate comingling of state-owned entities with commercial space companies. The key elements of the strategy were identified as follows:

- ". . . theft of intellectual property combined with a concerted and effective drive to create organic, national expertise across key space science and technology areas;
- direct integration of state-owned corporations and their technologies with commercial, space startupcompanies;
- penetration of American companies to obtain and further exploit U.S. technology or to influence those companies in a direction that serves China's domestic space priorities;
- investment in the U.S. space industrial base via front companies and multi-level off-shore accounts to facilitate early venture technology surveillance, infrastructure access and control of developing space capabilities and intellectual property;

[16] U.S.-China Economic and Security Review Commission (2019). ''A hearing on China in space: A strategic competition? 25 April. *https://www.uscc.gov/sites/default/files/2019-10/April%2025%202019%20Hearing%20Transcript.pdf.*

[17] U.S.-China Economic and Security Review Commission (2019). Report to Congress. *https://www.uscc.gov/sites/default/files/2019-11/2019%20Annual%20Report%20to%20Congress.pdf.*

- obtaining vertical control of the key space capabilities' supply chains or control of sufficient elements of those supply chains to influence space capabilities development in their favor;
- predatory pricing of space capabilities or elements of key space supply chains to control or dominate the market; and
- use of state-sponsored venture capital, finance, and market control mechanisms to surveille U.S. technology, interdependencies, business model innovations and other advanced concepts."[18]

Congress has responded to these activities most recently through the National Defense Authorization Act of 2020, which calls for the President, in consultation with the National Space Council, to develop ". . . a strategy to ensure the United States can effectively compete with other national space programs, maintain dominance in the emerging commercial space economy, and has market, regulatory, and other means available to address unfair competition from the PRC . . ." and also ". . . a plan to strengthen and streamline cooperation with international allies and partners in space."[19] However, additional action is needed to support the industrial base and by extension U.S. civil and commercial space activities it supports.

The Role of the Commercial Sector in Advancing U.S. Leadership and Global Relationships

As mentioned, several times, the primary strategic objective of NASA's activities in human exploration as seen from a national lens is geopolitical. However, that is not the only objective. In a previous appearance before the Aviation and Space Sub-committee of the Senate Commerce Committee in 2019, I opined that "United States leadership in space depends upon establishing a foundation that provides other nations and a nascent space-based economy with security and assurance regarding our national intentions and long-term commitment to aspire, inspire, and achieve—in short, to

---

[18] Cooley, T., Felt, E., and Butow, S. J. (2019). State of the space industrial base: Threats, challenges, and actions. 30 May. Air Force Research Laboratory and Defense Innovation Unit.*https://cdn2.hubspot.net/hubfs/4653168/AFRL_DIU_Report_State_of_Space_Ind_B ase_30May 2019_Final.pdf.*

[19] U.S. Congress (2019). National Defense Authorization Act of 2020, Pub. Law 116–92 (20 December). *https://www.congress.gov/116/plaws/publ92/PLAW-116publ92.pdf.*

lead."[20] My viewpoint has not changed; however, my sense of urgency has. Regarding low Earth orbit (LEO), which will be the focus of the rest of this testimony, NASA plans to transition activities now onboard the ISS to one or more commercial platforms in LEO, as well as some intended for deep space to the lunar orbiting outpost known as Gateway. In LEO, effecting this transition successfully requires closing the gap that now exists between public investment (NASA and the ISS National Lab) and revenue generation sufficiently to lead private investors to fund new orbital facilities for both public and private use. In turn, this requires hardware developers to create a portfolio of services attractive to a robust group of customers. In the case of Axiom, this includes first offering qualified customers the opportunity to fly to the ISS—establishing the procedures and agreements necessary to develop joint operations with NASA as well as conducting science and research— and later to Axiom's own modules while attached to the ISS. This in turn will lead to astronaut missions with a full complement of activities onboard Axiom station, on a commercial basis, once it detaches from the ISS.

To help visualize the latter, here is a quick video showing the assembly sequence: https://www.dropbox.com/s/z1s1paok8z8frax/Assembly%20Sequence_stills.mp4?dl=0.

Axiom's first private astronaut mission to the ISS is scheduled to fly in January of 2022, less than 6 months from now. Although he is not representing his country as a professional astronaut, one member of the first mission crew is an Israeli national and former Israeli Air Force fighter pilot who founded an investment firm that targets companies whose intention is to generate positive, measurable social and environmental impact alongside a competitive financial return. A close friend of Ilan Ramon, a member of the STS–107 crew who perished during the breakup of the shuttle Columbia in 2003, he will be conducting research onboard the ISS that is funded in part by the Ramon Foundation. A second member of the crew— which will be commanded by Michael Lopez-Alegria, a 4-time shuttle astronaut and former commander of the ISS—is a Canadian philanthropist and investor. Together with a third crew member and pilot who is American, all of the crew will be conducting research during their flight, bringing international payloads in through the ISS National Lab and opening the door to future missions with

---

[20] Dittmar, M. L. (2019). Testimony before the Senate Commerce, Science and Transportation Committee Subcommittee on Aviation and Space, 9 July (p. 2). "NASA's Exploration Plans—Where We've Been and Where We're Going." *https://www.commerce.senate.gov/services/files/19109A37-22BF-4268-9E5C-553880A435BE.*

professional astronauts and researchers from other countries. Other commercial companies such as Nanoracks, Made In Space/Redwire, and others have also developed their customer portfolios to include international customers.

NASA, and by extension the U.S. space companies serve as a vehicle of U.S. "soft power," shaping favorable behavior and outcomes among foreign entities, customers, and over time, competitors who do not operate according to American laws and values. American companies understand, rely upon, and promote protection of intellectual property, fair exchange, mutual respect, and enhancement of customer experience. Commercial partnerships with nations, institutes and firms abroad can also provide a means for countries to "bootstrap" into a human spaceflight or space science program, working with American companies to develop science and engineering initiatives supporting space activities, establishing long-term relationships. In the case of the ISS, increasing the number and duration of these relationships strengthens the viability of commercial companies offering services in LEO, enabling NASA to eventually transition off the ISS and turn its full attention to deep space with the assurance that its needs and those of its partners will continue to be met. It should be noted that this is not a "new" phenomenon; American businesses have played a crucial role in establishing relationships with other countries in a variety of sectors—not just space—and have frequently engaged in diplomatic initiatives and helped to deepen international ties. In space, all this proceeds from existing partnerships with NASA and is expanding to incorporate other sectors and customer bases over time.

The long-standing partnership between the government and commerce to promote American values and engage with other nations rests upon certain principles, however, including fair competition. At present, both Russia and China are competing with the United States by subsidizing launch costs and access to the Russian segment of the ISS and to Tiangong, the Chinese space station, respectively. Recently, a U.S. commercial company, Nanoracks, revealed that it had lost a commercial customer to the Chinese station for the first time.[21]. These developments threaten revenue sources for commercial space companies as national programs enter the commercial market with disruptive effect. Mechanisms to address unfair practices exist and have been used effectively in the past (for example, the 1993 agreement between the

[21] Foust, J. (2021). China's space station emerges as competitor to commercial ventures. Space News, 5 August. *https://spacenews.com/chinas-space-station-emerges-as-competitor-to-commercial-ventures/.*

Russian Federation and the U.S. regarding commercial launch services, which enabled the Russians to enter the market without upsetting normal competition),[22] however they must be prioritized by the U.S. government that recognizes the impact of such practices to the growing space economy and to America's leadership in space.

The PRC certainly recognizes the power of utilizing space activities—especially human spaceflight—to enhance national prestige and deploy soft power. To this end the Chinese have recently announced plans to fly 1,000 payloads on the Tiangong as it continues construction of the station.[23] In 2018, the PRC announced an agreement with the United Nations to "develop the space capabilities of United Nations Member States via opportunities onboard the CSS" (China Space Station).[24] Since that time nine experiments have been selected, with more in the pipeline.

Constancy of Purpose: Clear Policy Objectives and Strategy Should Drive Acquisition Approach

In LEO, NASA, and by extension its commercial partners and emerging companies aspiring to operate in orbit, are facing uncertainty. Plans to transition off the ISS at the end of its life requires clearly stated objectives, realistic timelines with milestones, and a firm commitment that has not been adequately communicated to date. In Axiom's case, the competitively sourced award that provides the company with exclusive access to begin attaching modules to the ISS by 2024 opens the door to a phased transition of U.S. government payloads, commercial or international payloads to Axiom once the modules are operational. However, Axiom's success— and that of other providers—would be greatly enhanced by a publicly released transition plan and "date certain" for ceasing operations on the ISS, developed in tandem with the international partners in the program. The date should be far enough in

---

[22] Office of the U.S. Trade Representative (published in the Federal Register, March, 1994). Text version of the Federal Register notice is available at: *https://www.govinfo.gov/content/pkg/FR-1994-03-10/html/94-5498.htm.*

[23] David, L. Can the U.S. and China cooperate in space? Scientific American, 2 August.*https://www.scientificamerican.com/article/can-the-u-s-and-china-cooperate-in-space/.*

[24] United Nations Office for Outer Space Affairs (2018). United Nations and China invite applications to conduct experiments on-board China's Space Station. *http://www.unoosa.org/oosa/ en/informationfor/media/2018-unis-os-496.html.*

the future that it enables NASA to complete currently understood work but not so far that it exacerbates existing uncertainty.

Any such plan should begin with clear objectives. At present, NASA is describing objectives in terms of hardware development in its new "Commercial LEO Destinations" program, adopting the same acquisition strategy that worked for its Commercial Orbital Transportation System (COTS) program and eventually for the Commercial Crew Program. There are several differences between the circumstances that gave rise to COTS over a decade ago, however, and those in low Earth orbit now. First, COTS was able to leverage an existing market for launch services that had developed and diversified over 50 years to include a substantial customer portfolio in both government and commercial sectors. Secondly, COTS was set up to provide cargo and later crew conveyance (under "Commercial Crew Program") that was well understood, the need for which was communicated early in the program.

In LEO, neither of these conditions obtain. While hardware development is well- understood, there is no diversified market to leverage, and NASA has yet to clearly define its needs after the ISS ends. The assumption that the same procurement approach that worked for COTS—in very different circumstances, where the market is just now emerging—is the best model for the CLD program, is questionable.

Furthermore, other than to ensure one or more platforms are available after the ISS ends, it is unclear what objectives NASA is trying to meet on behalf of the Nation. Clear policy direction is needed. For example, in my view an important national objective is to counter potential Chinese hegemony in low Earth orbit. If Congress would authorize LEO programs with this goal in mind, then NASA's acquisition approach should reflect this, using mechanisms that balance NASA's evolving role and needs in low Earth orbit with the risks, benefits, and management practices best suited to meet U.S. geopolitical interests.

At present, it is unclear that the current acquisition approach reflects a systematic strategy addressing higher order objectives—or indeed, what these objectives are. This is a critical point, because the consequences of these decisions have tremendous implications for the future of U.S. human spaceflight in low Earth orbit, and possibly beyond. Clear objectives and deliberate strategy must come first and should guide the acquisition approach—not the other way around.

A similar, systematic approach should be reflected in the next iteration of the ISS Transition Plan called for in the NASA Authorization and Transition

Act of 2017. It is well understood that any transition plan will be a work in progress, iterated upon as circumstances continue to evolve. However, objectives and a strategy for meeting them is every bit as critical as the tactical "transition" plan (of which acquisition is a part.)

The same can be said for resources. Whatever the objectives, strategy and tactical aspects of an ISS Transition Plan may be, adequate budgets available on a consistent basis are critical to meet objectives. The practice of relying on Continuing Resolutions (CRs) in lieu of regular order and budgets passed on time creates uncertainty and increased risk of failure. Unnecessary risk is beyond frustrating: It poses a threat to U.S. competitiveness and the ability to continue a permanent U.S. presence in LEO after the ISS program ends.

Whither Market Development in Space?

NASA is not an economic development agency. We should not expect it to act as such, either. The space agency's core competencies revolve around science, exploration, education, and technology as well as hardware development. If the U.S. government is serious about statements made in past years across the Congress to achieve a vibrant, commercial space sector operating in low Earth orbit and beyond, a conversation should be had about ensuring that appropriate economic goals and the means to achieve them are enshrined in policy. The responsibility to organize and foster growth of the commercial space sector should be housed within a designated place in our government that is appropriately resourced and staffed with the requisite expertise. The roles and experience of the Department of Commerce make it the obvious choice. Bearing in mind the sense of urgency, informed must be made in the immediate future that would help guide NASA regarding implementation, lest we risk falling behind China as well as other nations.

In LEO, hardware development is reasonably well understood. Both industry and NASA believe that the private sector is ready to develop next-generation orbital platforms. In Axiom's case this confidence is shared by capital markets. As mentioned at the beginning of this testimony, Axiom is developing its station entirely on investor funds and projected revenues from services, contracting with NASA to provide data and insight as the project proceeds. However, Axiom and any other company providing services in LEO, now or in the future, are dependent upon sufficient demand to close their business cases. If there is not sufficient, sustained, and growing demand, then the very companies the U.S. is depending on to assure U.S. presence—and soft power—in low Earth orbit, may fail. New entrants who bring innovation to the space ecosphere know that they must generate demand or fail. Suppliers who support the entire aerospace and defense sector rely upon sustained demand—indeed, this is one of the most important aspects of government programs. If suppliers fail, then the U.S. may have to look overseas for production. In the case where overseas production is critical to U.S. industry—as we are now seeing in the IT and telecom sectors—threats to U.S. security emerge. The same is true in space.

While governments cannot create markets, NASA can fund and conduct research that is too costly, risky or difficult for industry to undertake and provide the results to industry for use in engaging with customers. Businesses, in turn, are responsible for developing customers. In LEO, policy objectives that are beneficial to both the U.S. government and to industry—engagement with international countries and business, deployment of soft power by American industry, encouragement of new entrants into space technology and services, continued investment by capital markets, sustained U.S. presence in orbit around the Earth after the ISS ends—as well as other objectives I have not discussed such as using LEO platforms to create and enhance education and diversity programs to develop America's next generation of scientists and engineers—all depend on sufficient demand to sustain commercial platforms.

Actions the Government Can Take: Recommendations

- Utilize the regular order for negotiating and passing Appropriations Acts, avoiding the use of CR's and other stop-gap funding measures that increase uncertainty and negatively impact NASA's ability to pursue policy and programmatic goals.
- Establish clear policy goals for LEO (and beyond). As part of this, consider whether the time has come to make market development a legitimate aim of space policy, and whether the Department of Commerce might lend better focus and expertise to that effort. This discussion should consider the changing nature of NASA's role. When NASA can buy a service from the market, it will act only as a customer. When NASA must engage in development, it should do so with a very deliberate, objective-based strategy that may include not only whether a service is available, but whether purchasing from the market is in accord with national goals. In neither case, however, should NASA be expected to exercise oversight of the market.

- Formalize policy goals in another NASA Authorization Act. In the recent past, NASA Authorization Acts such as those in 2005, 2008, 2010 and the NASA Transition and Authorization Act of 2017 called for international and commercial partnerships in the exploration of space. "Constancy of purpose" as reflected in the Acts has been of great benefit in assuring program stability. Equally important, Authorization Acts send a signal to the international community of the intent of Congress to continue to assert U.S. leadership in space—a signal that is important not only to our allies, but to those who do not wish us well.
- Encourage NASA to deliver a transition plan for the ISS that establishes clear objectives, working milestones, schedule, clarity regarding its own needs beyond the life of the ISS, in accord with Section 303 of the National Aeronautics and Space Administration Transition Authorization Act of 2017.[25] In addition to the requirements specified by that Act for the plan, NASA should also discuss its role and rationale throughout the transition period in meeting national objectives (including geopolitical ones) as well as its own. Ideally, this rationale should drive, rather than be driven by any given acquisition approach.
- Recognize that both Russia and China are determined to achieve global leadership in space, and that the PRC has embarked upon a long-term, multi-tiered national strategy aimed at success. This strategy includes MCF and a long-term view of markets with deliberate efforts to undermine the U.S. industrial base and to "buy-into" U.S. entrepreneurial efforts for the purposes of transferring technology and/or expertise. Consider an interagency process, perhaps coordinated by the National Space Council or the National Security Council, to detect and deny such efforts early.
- Support U.S. commercial entities by adopting a "light touch" regulatory regime whenever possible.
- Shift U.S. Government investment in LEO to a more balanced approach that is less on hardware development and more on conducting research that may be useful in establishing demand while ensuring that U.S. commercial companies and their international and

---

[25] U.S. Congress (2017). National Aeronautics and Space Administration Transition Authorization Act of 2017 (Pub. Law 11510, Section 303). 21 March. *https://www.congress.gov/115/plaws/publ10/PLAW-115publ10.pdf.*

domestic customers have access to the ISS in the interval between now and ISS end of life (EOF). Specifically:

1. Reduce uncertainty (see above recommendation about an ISS transition plan) and avoid abrupt changes in policy that disrupt business plans and customer relationships. Recent examples of this include the NASA com mercialization policy, and more recently a review by NASA of the existing policy that enables American companies to fly payloads of their international customers via the ISS National Lab, which stopped payload processing for weeks. Such abrupt changes negatively impact the ability of companies to build demand for their services and undermine the role of American businesses in a competitive landscape (encouraging migration of those customers to Tiangong), thereby creating additional risks to the NASA plan to transition LEO to commercial providers, which is in turn the lynchpin of the U.S. government's policy objective to maintain U.S. presence in LEO in perpetuity.
   Fully utilize the ISS as a platform for commercial development during this interval, encouraging commercial research, development, and use across diverse sectors to facilitate the process of growing market demand. This should include areas such as entertainment, marketing, private and professional (national) astronaut missions. Commercial use should be balanced with the needs of researchers and government agencies, but with the recognition that the clock is ticking regarding development of demand needed to close the gap between government investment in public-private partnerships, investor funding, and revenue generation.

2. Consider establishing an interagency process to look at the economic development of low Earth orbit in a holistic way, focusing on how government money is spent to encourage and possibly incentivize innovation and entrepreneurship. "Hope is not a strategy;" rather, as a nation we must think and act deliberately regarding the intersection of markets and space policy.

Senator HICKENLOOPER. Thank you, Dr. Dittmar. Now, before I introduce Mr. Gold, I want to ask for unanimous consent to enter the following document into the record, an article from "Space News" authored by Mr. Gold, titled, "The Mind — "It is titled, "Mind the Gap in Low Earth Orbit." So, if we ask for unanimous consent, all in favor say aye.

[A chorus of ayes.]

Senator HICKENLOOPER. So be it. [The information referred to follows:]

COMMENTARY Mike Gold

MIND THE GAP IN LOW EARTH ORBIT

For more than 20 years, the International Space Station (ISS) has supported continuous international crewed operations. An entire generation has never known a world where people aren't living and working in space.

The success and longevity of the ISS is due in no small part to its international nature. Far too often, we take the ISS partnership for granted, when in fact it represents an unprecedented international collaboration that has proved that global human spaceflight cooperation is both possible and beneficial.

Moreover, for the past two decades, the ISS has proverbially and literally been an outpost on the frontier of science. More than 3,000 experiments have been conducted on the ISS in a diverse array of fields ranging from fundamental physics and Earth observation to biomedical studies and advanced manufacturing demonstrations.

Not only has the ISS supported unique science, the station has also been a singular source of soft power for the United States. America has served as the heart and glue for the station's international coalition, enabling the U.S. to maintain its role as the world's preeminent spacefaring nation.

However, in the harsh environment of space, nothing is permanent. Ideally, the ISS could serve as a platform for innovation for another decade and NASA should leverage every bit of utility it can get out of the station for as long as it can. Unfortunately, the ISS will inevitably need to be retired, and it is incumbent upon NASA and the entire American space community to avoid a space station gap.

This has happened before, with America losing the ability to launch astronauts into space and depending exclusively on Russia for crewed transportation for nearly a decade. This capability gap represented a holistic failure of planning and foresight that, thanks to public-private partnerships and commercial space capabilities, the nation is now starting to recover from. Yet, despite having only recently overcome a human space launch gap, the U.S. is already facing a space station gap that could be more pernicious than any other challenge that NASA has faced in the modern era.

THE PRICE OF INACTION

If the space industry and relevant policymakers fail to take robust action quickly, the U.S. will experience a crippling and lengthy space station gap. At a time when scientists are only beginning to understand the importance of microgravity research, development and manufacturing, the U.S. will lose access to low Earth orbit (LEO) and all the scientific benefits this unique environment has to offer. Additionally, America's ability to train astronauts and prepare for long-duration missions to the moon and Mars will be lost, and NASA's

SPACENEWS GRAPHIC

astronaut corps will shrink to a fraction of its current size.

Moreover, a space station gap will result in ceding both LEO and global leadership to China.

In April, China launched the first element of its new 66 metric ton three-module space station. China has learned from our experience with the ISS and is replicating both the capabilities of the station as well as the diplomatic and geopolitical benefits that a crewed LEO platform offers. During the next two years, China will conduct a flurry of activities to build this next-generation orbital outpost. Unless plans for an American commercial space station are executed with alacrity, this Chinese station will soon become the only crewed platform in LEO. In addition to ceding the benefits of microgravity research, development and manufacturing, as well as astronaut training to China, the U.S. will lose its role as the global leader in human spaceflight.

Space has always been closely intertwined with global politics. The Apollo era competition to land humans on the moon of course serves as the primary example of this, but the ISS's role in geopolitics potentially rivals even that of Apollo. Over the past two decades, the ISS has served as a beacon of global cooperation supporting experiments from more than 100 nations and astronauts from 18. The ISS has deepened existing relationships with longtime American allies while broadening U.S. outreach to countries that are new to human spaceflight. Additionally, under the auspices of American leadership, ISS operations, including launch activities, have been both transparent and safe.

If the U.S. cedes LEO to China, we will abandon not only science and astronautics, but also our values, leaving China to set the agenda for humanity's future in space. China will become the new global leader in astronautics and will enjoy the nontrivial political and diplomatic benefits that such a position brings. By becoming the only nation with a crewed platform in LEO, China will demonstrate, in both perception and reality, that its technologies and organizational capabilities are superior to the U.S. in this vital arena.

If the U.S. cedes LEO to China, we will abandon not only science and astronautics, but also our values, leaving China to set the agenda for humanity's future in space.

China already weaves space into its overall geopolitical strategy, including satellites and astronaut seats in its trade agreements and treaties. China also wisely leverages space in its relationship with the United Nations and has already selected payloads that will be flown on its space station from a variety of developing countries. A U.S. space station gap wherein China has the only crewed platform in orbit will further a narrative of American decline in the face of Chinese preeminence, the results of which will harm our nation's global soft power and the perception of the U.S. and the values it stands for.

Lastly, it's important to note the devastating effect a space station gap will have on the American space launch industry. America has spent billions of dollars developing new cargo and crew systems. These programs have energized the U.S. space launch industrial base and spurred innovations that have allowed America to recapture the commercial space launch market, which had previously been lost to overseas competition. Without a space station, demand for launch services will drop dramatically, jeopardizing not only American capabilities and science in LEO, but also U.S. launch capabilities and competitiveness. The negative effects of the loss of crewed LEO operations will thereby be felt far beyond NASA and will inevitably become a nontrivial threat to national security as well.

The overwhelming cost to the U.S. of a space station gap could not be more clear. Fortunately, the solution is equally apparent. Public-private partnerships and the rise of commercial space is transforming the space world.

Due in no small part to innovative leadership by NASA through initiatives such as the Commercial Orbital Transportation Services (COTS) program and the Commercial Resupply Services (CRS) contracts, the private sector has been able to deliver cargo and crew to LEO.

COMMERCIAL LEO DESTINATIONS

The obvious next step is for commercial firms to provide NASA with not just transportation, but a platform in LEO. A commercial space station could deliver all the benefits generated by the ISS along with renewed global outreach, enhanced affordability and continued scientific innovations. However, even with the private sector's efficiencies, designing, developing and deploying a commercial space station will require substantial time and resources.

The private sector can meet this challenge, but it cannot do so alone. As in the COTS and CRS programs, NASA has a critical role to play. Specifically, NASA must serve as a catalyst and customer for the new commercial station. The agency must do all that it can to help stimulate demand for microgravity services. This means supporting projects on the ISS now that demonstrate how the microgravity environment can be used to manufacture new substances, produce advanced technologies and develop cutting-edge medical treatments that could transform a variety of fields here on Earth.

NASA should focus its support on activities with the greatest near-term >

COMMENTARY Mike Gold

< > potential to generate revenue that could help pay for the costs of a future commercial space station.

Unfortunately, although important to both space station operations and American competitiveness, profitable on-orbit manufacturing is unlikely to develop in the near future. Therefore, NASA has an immediate and vital role to play as a sophisticated customer for a commercial space station, and the agency's Commercial LEO Destinations (CLD) project is a step in the right direction.

Over its 60-year history, NASA has been a trailblazer, not just in exploration, but also in technology investment and procurement methodology. Other Transaction Authority, which led to the rise and dominance of the American commercial space industry, was first utilized at NASA through Space Act Agreements (SAAs). SAAs were used for COTS, which garnered substantial private sector investment. Through COTS and other public-private partnerships such as the Bigelow Expandable Activity Module, NASA has expertly utilized its power as a customer to encourage outside investment. NASA purchasing systems and hardware sends a strong signal to investors and entrepreneurs who subsequently follow the agency's lead.

In regard to private sector capabilities, buoyed by private equity, SPACs and venture capital generally, the commercial space field has never enjoyed greater financial wherewithal. NASA would be wise to leverage this unprecedented period to further evolve the commercial space paradigm. Under COTS and CRS, the government paid for the private sector to develop services that the government then purchased. This model was wildly successful and saved the taxpayer billions of dollars. For activities that are within the reach of the private sector, NASA has and should continue to enter into public-private partnerships, allowing the agency to focus its resources on the frontiers of exploration and technology.

A Northrop Grumman Cygnus cargo tug developed under NASA's Commercial Orbital Transportation Services program approaches the International Space Station's Canadarm 2 in February.

NASA JOHNSON VIA FLICKR

Despite past success, the existing commercial space paradigm must continue to evolve, and the next step is for the government to serve primarily as a customer. The private sector has the desire and ability to fund the development and deployment of a commercial space station.

However, for companies to access sufficient capital to field a commercial space station, NASA must lean forward as much as it can to reassure industry that it will be a robust and ongoing customer. NASA's desire to continuously fly two crew members in LEO and perform 200 investigations per year is a good start. If NASA can incorporate such a commitment into the initial CLD SAAs and potentially include additional crew and investigations in the future, such actions will position the private sector for success.

Change is inevitable, but it is also beneficial and necessary for institutions, ideas and even individuals to survive and thrive. The environment in LEO is changing and American capabilities must also change to meet tomorrow's challenges. China is wisely investing in crewed LEO operations and making strong diplomatic overtures to build a global coalition for its space program. To avoid ceding LEO to China, the U.S. must lean into its traditional strengths, specifically, entrepreneurialism and innovation, both of which are driven by the freedom and diversity that are the twin pillars of American society.

Taking the concept of a commercial space station from idea to reality will require a concerted effort from all of us. NASA, Congress, the executive branch and industry must all work quickly and in unison. Together, we can build a bridge that will allow America to safely and successfully cross over the space station gap, on a path toward a future of peace and prosperity in LEO and beyond for the U.S. and all of humanity to enjoy. **SN**

MIKE GOLD IS REDWIRE'S EXECUTIVE VICE PRESIDENT OF CIVIL SPACE BUSINESS DEVELOPMENT AND EXTERNAL AFFAIRS

28 | SPACENEWS JUNE 2021

Senator HICKENLOOPER. Now, Dr.—or Mr. Gold, the floor is yours.

Mr. GOLD. Thank you, Senator.

Senator WICKER. Perhaps we could also enter the visual that Dr. Dittmar——

Senator HICKENLOOPER. Absolutely.

Senator WICKER. — was trying to show us.

Dr. DITTMAR. Thank you, Senator Wicker.

[The visual referred to was a video shown at the hearing.]

Statement of Mike Gold, Executive Vice President for Civil Space and External Affairs, Redwire Space

Mr. GOLD. Thank you, Senator. I will speak for both myself and Redwire. My CEO is right behind me, so he can kick me if I get out of line. So, I appreciate it. Again, gratitude to our Chair, Senator Hickenlooper. I would also like to express my appreciation to Ranking Member Lummis, to Committee Ranking Member, Senator Wicker, and Senator Moran, as well as my appreciation to the talented and intrepid staff that has helped support this hearing. And again, I want to say thank you for this extraordinarily important, timely topic of international collaboration and competition in space.

We often think of space as a vacuum. The word space itself can be interpreted to mean nothing. But I prefer to think of space as everything. Space is what connects the moon to the Earth and all of us to each other. The exploration of space unites this country and the world, in a way no other endeavor can. And the desire to unite humanity is at the very heart of the Artemis mission.

Through Artemis, NASA is assembling the largest, broadest, and most diverse international, beyond low Earth orbit, human space flight coalition in history. However, assembling this unprecedented international collaboration for Artemis was neither simple nor easy. Due to the failure of every single beyond LEO, American human space flight initiative to come to fruition since Apollo, there is great skepticism among both partner and rival nations, relative to NASA's ability to sustain a program to return astronauts to the surface of the moon.

This is why, more than any other rocket engine or piece of technology, bipartisanship is the key to the success of Artemis. Without robust and ongoing bipartisan support, we will not be able to lead a global coalition to the moon, Mars, or any other destination. We cannot unite the world, if we cannot first unite here in Congress. Which is why a bipartisan NASA authorization bill is urgently needed to both reassure partner nations, and send a message of unity, of purpose to our rivals.

I want to thank this committee for already passing NASA authorization language through the Senate, as part of the U.S. Innovation and Competition Act and urge your counterparts in the House to proceed with alacrity so that the entire Congress can adopt a NASA authorization bill that strongly and explicitly endorses Artemis.

Beyond technology that we developed the Artemis, the policy surrounding the program is just as important. America must lead in developing norms of behavior in space, to ensure that the exploration and development of the moon and Mars is conducted in a safe, sustainable, and peaceful fashion. While at NASA, it was the privilege of my lifetime to craft and lead the development and successful adoption of the Artemis Accords, which have now been signed by 12 countries. The principles of the Accords, such as transparency, interoperability, due regard, avoiding harmful interference, the sustainable use of space resources, and mitigating debris are the foundation that humanity's future in space should be built upon.

I hope that NASA continues to expand the family of nations that have adopted the Accords, and that the agency focuses, in particular, on bringing new partners from Africa and developing countries around the world, demonstrating that no matter how large or how modest their contributions are to Artemis, that all nations can join us in this unprecedented journey of discovery to the moon and Mars.

Closer to Earth, the International Space Station continues to represent the pinnacle of global cooperation in human space flight. We must send a clear and unequivocal message to both partners and rivals, that America will remain a leader in low Earth orbit, by extending the ISS through 2030. At Redwire Space, which is the only company to successfully manufacture items on the ISS, we are focused on leveraging the unique nature of the microgravity environment to develop new technologies and innovations that will help to bolster global communications that will heal the sick and that will feed the hungry.

Moreover, America must ensure that we do not suffer from a space station gap, which would cede the scientific, economic, and diplomatic benefits of crewed LEO operations to China. Therefore, in an upcoming authorization bill, we must provide robust support for the development and deployment, by the private sector, of a new commercial space station that will continue the legacy of the ISS, sustaining and expanding American international cooperation in low Earth orbit.

While opportunities for collaboration are many and varied, we must also not lose sight of the ongoing technological competitions in space that American can ill-afford to ignore. Specifically, the countries and companies that master orbital servicing, assembly, and manufacturing, will be the economic and national security leaders of the future. I am proud that, at Redwire, through the development of systems such as Archinaut, a satellite that literally assembles itself in orbit, we are contributing to American competitiveness in this vital arena.

However, much more needs to be done and the upcoming NASA authorization should include explicit support for Archinaut and public-private partnerships generally, that will result in even more ambitious technologies, leveraging deployable structures, robotics, and 3D printing. We can, and I believe, should engage with rival nations, including China, on norms of behavior in space, as well as benign, cooperative scientific activities, such as lunar sample swaps and sharing climate data. However, constant vigilance is the price we pay for liberty, which is why I hope the Senate will continue to urge NASA to redouble the agency support for public-private partnerships, which are the key to maintaining American competitiveness in space.

The journey of Artemis and NASA is to the Moon, Mars, and beyond. But if we properly balance collaboration and competition, the destination will be peace and prosperity.

Thank you for this opportunity to testify and I look forward to answering your questions.

[The prepared statement of Mr. Gold follows:]

Prepared Statement of Mike Gold, Executive Vice President for Civil Space and External Affairs, Redwire Space

I. Introduction

Chairwoman Cantwell, Ranking Member Wicker, Subcommittee Chairman Hickenlooper, Subcommittee Ranking Member Lummis, and distinguished Members of the Committee, I'm grateful to all of you as well as your intrepid staff for the opportunity to testify regarding the vital topic of international collaboration and competition in space.

All of us in the space industry are explorers. However, we're not just exploring the depths of space, we're also exploring new international relationships and innovative partnerships with the private sector. The benefits gained from such activities creates tremendous value and is a critical part of our Nation's journey into the final frontier.

In today's geopolitical environment, maintaining and growing our international relationships in space has never been more important. The space environment challenges us in a singularly harsh manner. Space is a crucible that demands the formation of global coalitions to bear the extraordinary difficulties and risks of exploration and development. International partnerships are not just a luxury but a necessity, and the U.S. is at an inflection point where our Nation's ability to lead future global coalitions will either be bolstered and revitalized or undermined and substantively damaged.

II. Sustaining the Continuity of Artemis

In regard to beyond low-Earth orbit (LEO) human space exploration, our Nation has, unfortunately, struggled to maintain its focus. Numerous beyond LEO human spaceflight programs have been initiated by NASA only to face termination by a subsequent Presidential Administration or Congress. The Vision for Space Exploration, the Constellation Program, and the Asteroid Retrieval Mission, are just a few of the initiatives our country engaged in which did not come to fruition. This failure to sustain a beyond LEO human spaceflight strategy has not gone unnoticed by our international partners. Our colleagues in Europe, Japan, Canada, and across the world have in many ways born the brunt of our national inability to execute on a proposed vision. This pattern of inconsistency has substantively damaged U.S. credibility on the global stage and has forced partner nations to question their relationship with the U.S. in space exploration.

As the Acting Associate Administrator for the Office of International and Interagency Relations ("OIIR"), I was often exposed to this growing skepticism of NASA's ability to sustain a beyond LEO human spaceflight program. My first overseas trip upon joining NASA in 2019 was to Paris to conduct negotiations for the European Space Agency's ("ESA's") contributions to the Gateway. My enthusiasm for international collaboration was met with justifiable skepticism from our European partners about whether the Gateway and Artemis would be sustained or, like previous NASA beyond LEO human spaceflight plans, would fade away and fail due to changing political priorities and/or a lack of funding.

Despite this skepticism, the OIIR team and our State Department counterparts managed to convince not just ESA's leadership, but the Japanese and Canadian governments that Artemis would be different. My primary argument was the bipartisan support that Artemis enjoyed. At the time I pointed to support from several Democratic policymakers, including the Chair of this Committee, Senator Cantwell, for the general goals and structure of Artemis. I personally reassured foreign officials that then NASA Administrator Jim Bridenstine was keenly aware of past history, and he was making every effort possible to build a bipartisan coalition which would sustain Artemis through any future changes in leadership in the White House or Congress. As a matter of fact, I was brought to NASA by former Administrator Bridenstine specifically to build a global coalition since he viewed such international partnerships as the key to sustaining the Artemis program.

Again, despite justifiable skepticism, our international partners agreed to join us and executed binding agreements in support of the Gateway. This support was manifested by the commitment of billions of dollars of international contributions toward building hardware for the Gateway and the Artemis program. Our international partners have embraced Artemis, appropriated substantial funding, and built their own plans around our strategy. In return, the U.S. cannot equivocate, turn back, or show anything less than the steadfast commitment that our international partners have already demonstrated.

Over the course of decades during which the U.S. struggled to sustain a beyond LEO human spaceflight program, the world has changed. China now represents a capable and attractive partner which is courting even our closest allies in Europe, Canada, and around the globe. In stark contrast to NASA, the Chinese space program has enjoyed unwavering political direction and consistency relative to their lunar plans. In the eyes of many, China represents a reliable alternative partner to the U.S. that does not suffer from the vicissitudes of the American political system and is rapidly growing its space-related capabilities and expertise.

Therefore, clear and explicit support for the Artemis program has never been more vital. The U.S. can no longer afford to change direction and must continue to embrace the Artemis program in a bipartisan fashion to preserve American credibility and our Nation's ability to assemble and lead the global coalitions that are necessary to succeed in space exploration.

For all of these reasons, I applaud the Biden administration for maintaining the continuity of the Artemis program with enthusiasm and alacrity. It was an honor to serve with then Acting Chief of Staff and Senior White House Appointee, Bhavya Lal, whose outstanding leadership allowed the Agency generally and Artemis specifically to emerge from the transition, which are always difficult periods, with clarity and constancy of purpose. Moreover, your former colleague and now NASA Administrator Bill Nelson has provided strong and consistent leadership and enjoys a close relationship with the President as well as the Congress which is vital to successfully implementing Artemis.

I would also be remiss if I did not commend the bipartisan leadership of Jim Bridenstine, including reaching across the aisle to place our current Administrator on the NASA Advisory Council. Administrator Bridenstine's unflagging efforts to build a broad and deep political coalition helped to ensure the future of Artemis. I will never forget when, after participating in a Women's Equality Day event at NASA Ames, Speaker Pelosi stated her support for Artemis by instructing the Administrator to work on getting the first woman to the Moon as quickly as possible. Subsequently, Vice President Pence tweeted about the Speaker's message thanking her for the support. Such alignment between two political rivals is all too rare, and kudos to Jim Bridenstine for his efforts to bring disparate leaders together in support of the Artemis program.

Of course, this Committee and its staff has consistently served as a shining example of bipartisanship. In my various private sector positions and while at NASA, I was always grateful for the vision and unparalleled bipartisan nature of the Commerce Committee's space policies. Under both Democratic and Republican control, including during Senator Nelson's tenure, this Committee has represented the best of Capitol Hill, always striving to find common ground and bridging divides.

Now, more than ever, we need bipartisan leadership, which is why Congress passing a NASA Authorization Bill is of paramount importance. As I described previously, it's vital for the U.S. to present a unified vision to both allies and rivals. The best way to accomplish this is through a bipartisan NASA Authorization which includes strong and explicit support for the Artemis program. I want to thank and congratulate this Committee for crafting a NASA Authorization as part of the U.S. Innovation and Competition Act (USICA) of 2021 which has been passed by the Senate. It's critical for the House of Representatives to also take action and for Congress to adopt a reconciled bill as expeditiously as possible. Both partner and rival nations have a sophisticated understanding of the Congressional process. The lack of a bipartisan NASA Authorization bill sends a negative message regarding America's ability to move forward with a unified vision for space exploration. Therefore, this issue must be rectified quickly before it leads to inevitable doubts and concerns that will damage America's ability to execute the critical global partnerships that are such an important part of the Artemis program.

III. Avoiding a Space Station Gap

Another area where policy consistency is critical is in LEO. The venerable ISS, which has now supported a continuous crewed presence for over twenty years, stands out as a dramatic success for American global leadership and policy continuity. Thanks to the ISS, the U.S. has been a hub for an unprecedented partner ship that defines the present and will influence the future of human space exploration. An entire generation has been born and grown to adulthood never knowing a world where there isn't an international team of scientists living and working on a space station. I believe that far too often we take this extraordinary accomplishment for granted. Having negotiated the binding Memorandums of Understanding for the Gateway which were based on the ISS's Intergovernmental Agreement, I have a great personal appreciation for the difficulty of assembling the ISS coalition as well as maintaining it. I have always felt that the ISS should receive a Nobel Peace Prize acknowledging the herculean work of those who transformed the station from dream into reality.

Yet again, I would like to thank the Committee for including in the NASA Authorization Act, which passed the Senate, language extending the ISS to at least 2030. The ISS still remains the pinnacle for human spaceflight and we should leverage the station's singular capabilities for as long as possible. Again, Congress adopting an authorization bill extending the ISS will be extraordinarily important to provide our international partners with certainty relative to the future of the station.

Moreover, per the commendable authorization language passed by the Senate, America's goal is to maintain a continuous human presence in LEO. Extending the ISS is an important part of achieving this objective, but of equal importance is ensuring that when the ISS is retired, there is a commercial space station that will carry on its legacy. The ISS itself took decades to construct and America now faces the troubling prospect of a space station gap. The U.S. can ill afford to lose its presence in LEO. Astronautics opportunities, which are the grist for the mill of inte national collaborations, will always be more plentiful in LEO than for beyond LEO operations. Additionally, we're only now beginning to realize the incredible potential of microgravity manufacturing, research, and development. The microgravity environment represents an entirely new arena for trailblazing scientific and commercial endeavors. What we will learn from and what we can do in microgravity has the potential to revolutionize fields as diverse as drug treatments and medicine to communications, agriculture, and construction.

America cannot lose access to this unique environment at exactly the time when others are beginning to establish a foothold in LEO. Specifically, over the course of the past year, China has launched its own space station which has now received both resupply missions and astronauts. Only a few days ago, China launched their most recent crew to their station, beginning the longest duration mission Chinese astronauts have ever undertaken. Moreover, China is aggressively courting our existing ISS partners, particularly in Europe, to join their new crewed LEO operations. Again, for two decades the ISS has allowed the U.S. to serve as a hub for global astronautics activities. As the world's leading Democracy, we must not cede the substantial diplomatic and political soft power benefits of crewed space station operations to China or any other nation.

Moreover, a space station gap would have a dramatic negative impact on American launch capabilities. Losing the demand for commercial cargo launches to the ISS as well as commercial crew (not long after investing billions of dollars to develop this capability), would impede gaining experience with these systems as well as continued innovation and growth. The damage done to American launch capabilities and international competitiveness would hurt our overall industrial base creating issues not just for civil space operations but for national security space activities as well.

Avoiding a space station gap will not be simple or easy. Developing and deploying a new space station is a nontrivial effort that could take a decade or more. NASA is wisely turning to the private sector to leverage commercial innovation and efficiencies to build and operate this new station. I'm excited to see NASA's Commercial LEO Destinations (''CLD'') program progress taking us a step closer to ensuring that the U.S. and our partners maintain a continuous human presence in LEO. At Redwire Space, we are the only organization that has ever successfully printed objects on the ISS and we are a global leader in microgravity manufacturing and R&D. We will apply our company's singular capabilities and experiences with micro-gravity manufacturing, along with our innovative roll out solar arrays, robotics, digital engineering, sun sensors, star trackers, and other technologies to ensure the success of the CLD effort.

Finally, I want to applaud this Committee for passing language that reinforces America's commitment to maintaining a continuous presence in LEO through the development of a commercial space station. I hope that the House moves forward quickly with a NASA Authorization Bill which includes identical language allowing Congress to send an unequivocal message to partners and allies alike that we have learned from history and that this Nation stands united in support of preventing a space station gap.

IV. Maintaining Technological Leadership Through Public-Private Partnerships

There can be no question that we are in a great powers struggle with China. This rivalry extends to space where the implications of the U.S. falling behind China technologically would have devastating consequences. This is a race that the U.S. cannot afford to lose, however, in many critical areas we are already facing the prospect of falling behind. Specifically, the U.S. must redouble its efforts to support next-generation space-based robotics. We are entering a new era where satellites and robotic systems are merging, and America must embrace this change and excel at these advanced technologies.

For example, at Redwire Space we are developing systems such as Archinaut (also referred to as 'OSAM–2'), a satellite that, after deployment, will use robotics to build itself. In addition to robotic assembly, we are developing satellites that leverage 3D printing to create far larger objects in space than could have ever been placed into a rocket fairing. These space platforms that construct themselves after being deployed in space will offer substantially enhanced capabilities as well as resilience to attack that could never be achieved by traditional satellite systems. While Redwire Space has unparalleled heritage and expertise in this field, China is well aware of the importance of space-based robotics and has already deployed Shijian- 17, a satellite with, among other features, a robotic arm that could be used to grapple satellites. Shijian-17 has already flown unusual orbital maneuvers bringing it closer to other satellites generating suspicion and concern. Again, the U.S. must bolster its efforts to equal and exceed Chinese space-based robotic capabilities to protect our industrial competitive edge as well as the Nation's overall economy and security.

As noted previously, China's autocratic regime does have the benefit of policy consistency and the Nation has focused on developing world-class space technology for decades. China's efforts are coming to fruition and the U.S. faces an unprecedented challenge to maintain space supremacy or, in the long run, even parity.

The best and possibly only way for the U.S. to meet this existential threat is for America to embrace its great strengths that China struggles to or cannot reproduce. Specifically, the U.S. must redouble its efforts to leverage

American entrepreneurialism. A free society, grounded in the rule of law, with a diverse population, will always innovate at a greater rate than more homogeneous nations where freedoms are severely curtailed. America created the commercial space revolution, and our private space sector is the envy of the world. Programs such as NASA's Commercial Orbital Transportation Services ("COTS") and the Commercial Resupply Services ("CRS") contracts, played a critical role in creating the commercial space industry as we know it today.

In order to effectively compete with China, the Congress, NASA, the Department of Defense, and the Intelligence Community, must all embrace public-private partnerships to an even greater degree than what is occurring today. The government must support commercial development, playing the role of both catalyst and customer for innovative technologies. Commercial space represents a force multiplier that has already transformed the nature of the industry. However, there is much more that can be done by the U.S. government to make efforts such as COTS and CRS the rule rather than the exception. Again, whenever possible, the government should serve as a customer for innovative capabilities that will maintain and accelerate U.S. leadership, particularly in critical areas such as orbital servicing, assembly, and manufacturing. Only by substantially expanding the use of public-private partnerships can the U.S. effectively protect itself and the world from emerging threats that will only become more pernicious in the years to come.

For all of these reasons, the NASA authorization bill should include explicit support to bolster Archinaut specifically and on-orbit servicing, assembly, and manufacturing technologies generally. This is a critical area of technological endeavor that the U.S. must excel in and explicit direction via an authorization bill is key to maintaining American competitiveness.

V. Establishing Norms of Behavior in Space

It is vital for America to lead not only in technology but in policy as well. As in any new frontier, establishing rules and norms of behavior will play a critical role in preventing conflict and ensuring that space is developed in a safe, sustainable, and peaceful fashion. While at NASA, it was my privilege to craft and negotiate the Artemis Accords which, in less than one year since their inception, have now been signed by twelve nations.

The Artemis Accords were successful due in no small part to the unprecedented partnership between NASA and the Department of State, and I want to take this opportunity to thank Jonathan Margolis, Acting Deputy Assistant Secretary for Science, Space, and Health at the Bureau of Oceans and International Environmental and Scientific Affairs for his support, partnership, and friendship in developing the Accords. I also need to thank another friend and colleague, Gabriel Swiney, the Department of State's lead civil space attorney who was one of the first people I discussed the Accords with. Gabriel's vision and knowledge are reflected throughout the Accords and his skill made them a success. Finally, the Artemis Accords would of course not exist without the outstanding leadership of Jim Bridenstine, who provided strong and unflagging support for the Accords as well as the partnership with the Department of State, and the former Administrator simply wouldn't take no for an answer in his efforts to bring me to NASA.

The Artemis Accords present the model for future policy initiatives wherein NASA and State work in unison. Moreover, the Accords would not have succeeded without the coordination and assistance provided by the National Space Council, and its former Executive Secretary, Dr. Scott Pace. I have greatly benefited from Dr. Pace's support over the years, and his leadership was vital for both the Artemis Accords and the signing of the binding Gateway agreements, particularly the agreement with the Government of Japan. Finally, I want to again thank the Biden administration's leadership including Secretary of State Antony Blinken, National Security Advisor Jake Sullivan, NASA Administrator Bill Nelson, NASA Deputy Administrator Pam Melroy, and Senior Advisor to the Administrator, Bhavya Lal, for their strong support of the Artemis Accords. Again, continuity is vital, and I am personally grateful for the Biden team's robust support for building international partnerships and establishing norms of behavior in space.

Due to the Accords, the Artemis program has established the largest and most diverse human space exploration coalition in history, but much more remains to be done. Specifically, more countries should be added to the Accords and the Artemis program. For example, I hope that NASA and the Department of State continue and accelerate efforts that I worked on during my tenure at OIIR to bring African nations into the family of the Artemis Accords. I believe that the benefits of space exploration and space-based capabilities are particularly important for developing nations and that the unique views and voices of African youth in particular can greatly contribute to the Artemis program. Moreover, China has targeted Africa both economically and politically, and NASA should not cede African partnerships to rival nations. Additionally, it would be beneficial for a few more of our traditional space allies, specifically, France and Germany, to join the Accords demonstrating unified support in Europe for responsible norms of behavior in space. Finally, I hope that efforts at NASA and State are continuing for India to join the Accords. India's robust and growing space capabilities make their support for norms of behavior critical to achieve a peaceful and prosperous future in space for all of humanity to enjoy.

The Accords establish a strong foundation for beyond LEO civil space activities, building a future based on transparency, interoperability, the full and public release of scientific data, avoiding harmful interference, and environmental sustainability. However, new rules and norms need to be established for national security operations. The Tenets of Responsible Behavior in Space, described in a July 7th Memorandum by the Secretary of Defense, represents a good start. However, establishing an international coalition to support these tenets is vital to their success and universal adoption. The Artemis Accords serve as a model for how such coalitions can be built, and I know that myself and others who were involved in developing and implementing the Accords stand ready to assist our colleagues at the Department of Defense and Space Force in their own efforts to establish vital norms of behavior in the national security realm.

In addition to explicitly authorizing the Artemis program, I hope that the upcoming NASA authorization bill includes language complimenting NASA's work in support of the Artemis Accords, signaling to both partners and rivals that the Congress strongly endorses establishing global norms of behavior to ensure a peaceful and prosperous future in space.

VI. Collaborating and Competing with China

For better or worse, the U.S.—Chinese relationship will determine the future of space exploration and development. As described previously, the U.S. cannot afford to fall behind China in critical capabilities such as space-based robotics, in-space manufacturing, and space-based solar power. However, there are nontrivial opportunities to collaborate with China that could benefit both countries and advance global prospects for peace and prosperity.

Specifically, NASA and the Department of State should continue to advocate for China to sign the Artemis Accords. Although China is not a part of the Artemis program, and the Accords were written to ensure that nations participating in Artemis abide by international treaty obligations and norms of behavior such as the full, free, and timely release of scientific information, China signing the Accords would send a strong signal of unity and global consensus relative to the simple and intuitive principles of the Accords. The Accords were a product of lengthy negotiations between eight different space agencies and ministries of foreign affairs. The substance and text of the Accords represents the common ground among countries with a diverse set of policy positions (e.g., the U.S. opposes the Moon Agreement whereas Australia is a signatory). The Accords were written to be as inclusive as possible and, in particular, the United Arab Emirates officials who participated in the drafting process were a strong voice for ensuring that the Accords could be signed by any nation with a desire to support peaceful space operations and development. The Artemis Accords were explicitly and exquisitely crafted to unite nations with disparate views, bringing the world together to prevent conflict and ensure harmonious activities on the Moon and Mars. China committing to the Accords would be a welcome development that I believe would benefit both nations and global space policy discourse generally.

During discussions at the United Nations Committee on the Peaceful Uses of Outer Space, the State Department has invited China and all nations to sign the Artemis Accords. Beyond such interactions at the United Nations, I recommend that the State Department continue the practice of convening Civil Space Dialogues with China, to discuss the Accords and norms of behavior in space. If China were to join the Artemis Accords, like with other signatories, it would provide specific, actionable commitments that the country would be held to for civil space operations on the Moon, asteroids, comets, and Mars.

I have high hopes that a productive conversation could be held with China on norms of behavior due to my experiences with several global efforts that included Chinese participation. For example, prior to joining NASA, I served as an industry member of The Hague International Space Resources Governance Working Group, which brought together experts from around the world to draft 'building blocks' in support of establishing rules for space resource exploration and utilization. One of China's leading law professors participated in The Hague Working Group and his input was robust, constructive, and productive. As a matter of fact, some of the text and ideas generated by The Hague Working Group were adopted as part of the Artemis Accords.

Moreover, I am currently supporting the efforts of the Global Expert Group on Sustainable Lunar Activities ("GEGSLA"), which is a NGO that, as the name indicates, is bringing together policy and legal experts from across the world to develop norms of behavior for the Moon. The GEGSLA involves many of the same academic, industry, and government experts as the Hague Group, including several Chinese nationals who have been working side-by-side with, among others, American commercial space leaders to establish rules of the road for lunar activities which are based upon not only the Outer Space Treaty of 1967, but the prior work done by the Hague Group and the Artemis Accords. While I know from personal experience that coming to agreement on even general principles for space exploration and development can be extremely difficult, I firmly believe that this is an area where the U.S. could and should engage with China in an attempt to identify common ground. At the next Civil Space Dialogue between the U.S. and China, in addition to norms of behavior, other forms of collaboration could be discussed. For example, a swap of lunar samples would be a low-risk means of initiating cooperation in space with China, which NASA could engage in while staying well within the bounds of the Wolf Amendment. Finally, the U.S. is already collaborating with China and other nations on the Space Geodesy Project, which produces a wide variety of information for Earth observation and climate science. Additional collaborations with China on climate research is another area which, if handled correctly, could benefit both nations without running afoul of the requirements of the Wolf Amendment.

VII. Conclusion

Collaboration and cooperation must both be harnessed to ensure that humanity's journey into the undiscovered country of space is safe, peaceful, and prosperous. Again, we cannot afford to fall behind in critical new areas of space technology and must maintain the continuity of the Artemis program as well as maintain a continuous crewed presence in LEO. However, we must also reach out to both allies and rivals to collaborate whenever possible to support norms of behavior in space and mutual areas of scientific interest and concern such as climatology.

As mentioned previously, Redwire is a global leader for in-space manufacturing. We have an expertise in building microgravity products that will enable ambitious space exploration missions and improve life on Earth. Redwire, and all of us in the private sector, must join with our colleagues in government to build not just tech nology, but the future. This is an in-space manufacturing endeavor that will demand unprecedented work, coordination, and support from across industry and government. Although we will face many challenges, I remain confident in America's ability to build a new era in space of peace, freedom, and prosperity, benefiting the world and many future generations to come.

Senator HICKENLOOPER. Thank you. And you did not even mention the companies that Redwire has in Colorado.

Mr. GOLD. I will say, since it is just between us, they are three of our best.

[Laughter.]

Senator HICKENLOOPER. All right. Now, we are ready to hear from Dr. Sanders.

Statement of Dr. Patricia Sanders, Chair, Aerospace Safety Advisory Panel, National Aeronautics and Space Administration

Dr. SANDERS. Thank you, Senator Hickenlooper, Senator Lummis, and Wicker, and Members of the Subcommittee. I thank you for the opportunity to appear before you today to discuss NASA's international collaboration and competition in space.

The panel that I chair has the serious charge to provide both NASA and the Congress with advice for the intelligent management of the risks that are inherent in human space flight. NASA has been at the forefront of human space flight for decades, formulating the missions, defining the requirements, executing the programs, performing management integration, and leading launch processing and mission operations.

NASA still leads in human space exploration, but the Agency's role is evolving with critical implications for safety and risk management. The Agency is not the same as it was 10 years ago and will not be the same in even another 5 years. The rapid growth of commercial space services, and increasing global interest in space, have changed the environment and the landscape will not return where NASA is the only, or even the major, actor. This has tremendous upside potential and equally tremendous challenges for managing the risk of human space exploration. At the same time, the exploration endeavors NASA leads are ever more complex and have ever more risk involved.

NASA has been gradually and tactfully and successfully adjusting to a changing role and set of responsibilities, as it shifts from principally executing its programs and missions to commercially and internationally acquiring significant key elements and services. And our Panel firmly believes that NASA must now take a more strategic examination of the Agency's evolution in the emerging environment, while continuing to manage a safe and complex human exploration campaign.

NASA's challenge is the melding of traditional and innovative approaches, including significant systems engineering and integration complexities, and the certification of commercial human space flight capabilities with high levels of risk. There are clear advantages to leveraging industry innovators and international partners, but NASA must still manage and be responsible for the overall risks, even when the Agency neither controls nor dictates the materialNASA, first, must determine how to exercise appropriate accountability, or how to hold its vendors accountable, for the safe and successful accomplishment of its mission.

Second, we believe it is imperative to define the overall architecture for the highly complex Artemis mission. The Agency should identify how each individual element, regardless of provider, fits the architecture, and the top-level requirements for each element to fulfill its necessary function in the overall structure. This should allow NASA to focus on the right set of priorities at the right time and to communicate expectations to all the contributors: internal, commercial, and international, in a consistent manner.

Third, all of this is complicated by the Nation's current lack of a comprehensive regulatory framework for human commercial spaceflight. NASA retains full accountability for its missions, but no external government regulations or standards exist to set a baseline level of expectation for the provider related to human safety. The few existing regulatory pieces leave a gap related to human on-orbit safety, and the space industry that impacts human safety. In particular, as has already been mentioned, there is an immediate and compelling need to designate a civil agency to oversee and coordinate space traffic management.

Given the importance of space to international security, technological leadership, and international competitiveness, it is vital for the Nation to act now to preserve the safety of space operations and their environment.

In closing, I note some consistent advice themes from the Panel. First, it is a need for a constancy of purpose, as mentioned, sustained commitment, and a clear understanding of the objectives. Second, is the importance of setting challenging but achievable schedules, and not allowing undue schedule pressure to lead to decisions adversely impacting safety and mission assurance. And third, is technical baselines and schedules that are mutually consistent, realistic, and achievable, and supported by adequate and stable resources.

We encourage NASA, in partnership with the Congress, to hold fast to the foundations of risk management, while embracing and not fearing alternative methods to achieving those fundamentals.

And I look forward to your questions. Thank you. [The prepared statement of Dr. Sanders follows:]

Prepared Statement of Dr. Patricia Sanders, Chair, Aerospace Safety and Advisory Panel, National Aeronautics and Space Administration

Senator Hickenlooper and Members of the Subcommittee, thank you for the opportunity to appear before you today to discuss NASA's international collaboration and competition in space.

As you know, the Aerospace Safety Advisory Panel is charged with advising both the NASA Administrator and the Congress with respect to the safety and risk of human space flight as well as other safety related matters at the Agency. In opening, I would like to emphasize that the Panel feels the responsibility to provide advice that promotes the driving down of risk to the lowest reasonable level consistent with accomplishing the mission. Space exploration is inherently dangerous; the environment is hostile, and the systems needed to survive in it are complex. Our charge is not to avoid any and all risks, but to provide advice and feedback for the intelligent management of those risks.

NASA has been at the forefront of human space flight for decades, and for much of that time it executed the programs, formulated the missions, defined the requirements, and performed management integration of all the elements composing the solutions for all of the components. To do this, we believe that system. NASA personnel performed the engineering analyses, and they led launch processing and mission operations.

NASA leadership in human space exploration is still preeminent, but the Agency's role is evolving with critical implications for how risk and safety will be managed in the future. The Agency is not the same as it was ten years ago, and most assuredly, it will not be the same in another ten years—even five years—from now. With the rapid growth of available commercial space services, and increasing global interest in space, the environment in which NASA operates has changed; NASA will not return to a landscape in which it is the only, or even the major, actor. These developments have tremendous upside potential—and are accompanied by equally tremendous challenges for managing the risk of human space exploration. Concurrently, the human exploration endeavors NASA is leading are becoming ever more complex, and with more risk, from the lunar exploration to eventual excursions to Mars and beyond.

Over the past several years, NASA has been adjusting to a changing role and set of responsibilities as it shifts from principally executing its programs and missions to commercially acquiring significant key elements and services. The Agency has gradually and tactically adapted and succeeded in meeting challenges as they arise. Regardless of their tactical achievements to date, the Panel firmly believes that it is critical at this time that NASA take more strategic scrutiny of the role the Agency should undertake going forward. How the Agency plans to evolve and transition to an organization that more frequently procures human space flight capabilities as services, while managing a safe and wholly new human exploration campaign, is a key strategic question that has the Panel's attention. We continue to emphasize the importance for NASA to strategically define its mission, its guiding principles, and its vision for the Agency's leadership role in the future in order to ensure that risk is managed appropriately.

The emerging challenges for NASA involve the melding of traditional and innovative program approaches, including the significant systems engineering and integration complexities, and the certification of commercial human space flight capabilities that carry high levels of risk. The Panel has noted clear advantages to leveraging the industry innovators, but NASA must still manage and be responsible for the overall risks, even when the Agency does not control nor dictate the material solutions for some of the campaign components and services, such as the Human Landing System. It is critical for NASA to be able to manage the integrated risk and achieve the right balance with its providers.

To do this, first, the Panel believes that NASA needs to figure out how to exercise appropriate accountability—or how to hold its vendors accountable—for the safe and successful accomplishment of its mission across the full spectrum of acquisition and development approaches. As the breadth and types of relationships develop, expand, and become more complex to achieve NASA's mission safely, and with good understanding of the risk involved, it is critical for the Agency to have and to use the appropriate tools including acquisition processes and contractual structures. There is not a one-size-fits-all approach anymore, and having an overarching view of what the Agency is trying to achieve should lead to a flexible and thoughtful deployment of the tools in the toolbox.

Secondly, the Panel believes it is imperative to define the overall architecture for the highly complex Artemis mission sets. The Agency should identify how each individual element—regardless of provider—fits the architecture, and define the top-level requirements that must be met in order to for the element to fulfill its necessary function in the overall mission structure. This work can then form the foundation for the system engineering and integration. The complexity of the Artemis ecosystem, along with the expected evolution of requirements—which involves creating and maintaining an architecture that can be updated, adjusted, and can incorporate the latest innovation or new technology—can more effectively be managed by an integrated approach best achieved in a program construct. This should allow NASA as a whole to focus on the right set of priorities at the right time and to communicate expectations to all the contributors—internal, commercial, and international—in a consistent manner. In addition, all players understanding enterprise-level requirements, organized in a program construct, early in the process helps to identify opportunities and areas to pursue open architecture paradigms and reduce expensive, complicated, and bureaucratically burdensome design and contractual changes later.

Thirdly, I note that NASA's approach is complicated by the Nation's current lack of a comprehensive regulatory framework for human commercial spaceflight. Presently, NASA retains full accountability, but no external government regulations or rules exist, which may help the Agency manage risk, or even set a baseline level of expectation for the provider, related to human occupant safety. There are some regulatory pieces in place. At the highest level is the Outer Space Treaty. The Federal Aviation Administration is responsible for licensing commercial launches and reentries, with a specific focus on the safety of the uninvolved public on the ground. The Federal Communications Commission is responsible for licensing radio broadcasts from space. The National Oceanic and Atmospheric Administration is responsible for licensing remote sensing operations. NASA and the Department of Defense are key players in space, but they are not regulatory agencies. That leaves a gap in authority specifically related to on-orbit safety, both for humans and the management of an increasingly more active satellite industry that will eventually impact human safety. Given the importance of space to national security, technological leadership and international competitiveness, our Panel believes it is vital for the United States to act now to preserve the safety of space operations and the safety of the environment.

Consequently, I would be remiss if I did not repeat a standing Panel recommendation to the Congress. We feel very strongly that there is an immediate and compelling need to designate a civil agency to oversee and coordinate space traffic management. NASA, lacking any other framework, has established guidelines and standards for space traffic management, but there must be leadership and coordination at the national level.

Finally, I would like to reiterate some consistent advice themes from the Panel:

- First, we have consistently maintained that mission success requires a constancy of purpose, a sustained commitment, and a clear understanding of objectives.
- Second, a key issue, repeated year after year, is the importance of setting challenging but achievable schedules, and not allowing undue schedule pressure to lead to decisions that adversely impact safety and mission assurance.
- Third, it is important to establish technical baselines and schedules that are mutually consistent, realistic, and achievable—supported by adequate and stable resources.
- And, we have continuously maintained that while NASA should never lose sight of the fundamentals in risk management for successful program execution, there is no one approach that dictates success, and there should be an openness to learning and accepting alternative means to understanding and controlling margins.

So, as NASA continues its deep space exploration, we encourage the Agency, in partnership with the Congress, to hold fast to the foundational standards of risk management while embracing new approaches and not fear alternative methodologies to achieving those fundamentals.

Thank you. I look forward to your questions.

Senator HICKENLOOPER. Great. Well, thank you all for your— your comments and appreciate the—the breadth of experience you bring to the table.

We will now start a round of questions. Each of us will get—well, hopefully, will get 5 minutes. We try to stay on schedule. It is always hard. I always ask if, in responding to a question, if you can keep it to a minute, that allows us to get four questions in. If necessary, we can come back to a second round of questions or—there are so many questions to ask each of you. Anyway, I will start.

I will start with Mr. Gold. Appreciate your work at NASA with the international partners and establishing these—the Artemis Accords and that cooperative spirit. In your testimony, I looked at it last night, you said "space is a crucible that demands the formation of global coalitions." So, how can Congress support NASA's goals to expand the coalition of international partners in space?

Mr. GOLD. Thank you, Senator. I can tell you, and I am sure the former Administrator would agree, NASA pays a lot of attention to Congress. And the rhetoric, you know, obviously, legislation that has passed, it determines whether NASA is aggressive in going out and forming international coalitions or holds back and really what the Agency does.

So, I think it is extremely important, again, in an authorization bill, that there be explicit encouragement to the Agency, both to expand our international cooperation, generally, as well as specifically to expand, again, the family of the Artemis Accords. That there are so many countries that need to be involved that the Agency should be aggressively reaching out for, we need to lead by example and establish norms of behavior before there is conflict in space. And I know it is cliche′, an ounce of prevention now, will prevent a pound of trouble down the road. So, including even a sense of Congress, about the Accords, international growth, and being aggressive to establish norms of behavior, I think, would be extremely helpful to assure a peaceful future.

You know, other countries may or may not, you know, agree to the same norms that we do, but if we lead by example, we can take pressure there and create a coalition of willing.

Senator HICKENLOOPER. Absolutely. That is—I think you are spot on. I could not agree more.

Mr. GOLD. Absolutely.

Senator HICKENLOOPER. But I want to take—and Mr. Gold talked a little bit about this, the implications of a—of a gap. And I guess, Dr. Dittmar, let me ask you, where do you see the—what are the implications if we do end up with a space station gap, where the United States is not taking a leadership position?

Dr. DITTMAR. So, as—thank you for the question, Senator. As has already been pointed out, China has a space station that is flying in low-Earth orbit. This is actually their first module of a station that they are going to build out. And China is leveraging that space station. They have formed partnerships with UNOOSA. They have already—they recently announced that they had agreements to fly over 1,000 payloads. We have had American companies that have already said that they are beginning to lose customers to China and the Chinese station, because the Chinese are using state authority to subsidize access to station for commercial customers, as well.

And so, if what we are saying is that we want to have a follow- on station that is privately—a public-private partnership, or is owned by a private entity, that entity absolutely depends upon access to customers that are not being, in effect, stolen away by a state agency, that is doing that, you know, intentionally to, sort of, undermine that capability.

The other thing, frankly, is that, you know, robust presence in low Earth orbit, for Americans, has been a part of policy, whether stated or not, in the United States for more than 20 years—actually longer than that and—going back to the origin of NASA. And when you look toward an authorization bill, and I think it is really important to foot stomp, that U.S. presence in low Earth orbit is expected in perpetuity, that it is the policy of the United States that we will continue to do so, through whatever means. However, I think—and low Earth orbit is, frankly, our foothold going out into outer space. We need continuity of human space flight and human activity, beginning on the ground, through low Earth orbit, into deep space. And that chain is very, very important to American security, to American business, to American research and science, to American technology development and we cannot afford a gap.

Senator HICKENLOOPER. Yes. Well, we agree on that. Dr. Sanders, the GAO has expressed concern on the maturity of advanced lunar spacesuits. It is a classic case. But I think a lot of what all— everyone has talked about, and especially Mr. Bridenstine, the safety of these missions is—there are risks from 100 different directions. And can—I was going to ask you to describe how you believe that—that robust Congressional funding—that continuity of funding and authorizations could support—help NASA support the goals of safety and making sure that, you know, mission schedules can end up impacting safety and assurance.

Dr. SANDERS. Well, we have been a consistent proponent of adequately resourcing all of the complex requirements necessary for safely executing. When resources are not adequate, or not stable, the—NASA and their decision-making process on designs, end up making design choices that may be are premature, maybe they are—take risks that they would not necessarily take if they did not have to live within a resource that is not adequate, or they do not have the stability of the resources, it forces you to make choice sooner than you might want to take. It forces you to sometimes, down select earlier than you understand all the implications of competing designs. And so, it is very important to have the adequate resources and to have them in a stable fashion.

Senator HICKENLOOPER. Right. Great, I appreciate that. I am already out of time. I cannot believe how fast this goes by. But I am not done with all of you. Mr. Bridenstine, you will get your moment.

I am going to turn it over now to the Committee Chair, Senator Cantwell, who has been working on these issues for a long time, as long as—probably longer than most of you—most of us. Anyway, let me turn it over to the Chair.

Statement of Hon. Maria Cantwell, U.S. Senator from Washington

The CHAIR. Thank you, Chair Hickenlooper, and thanks for holding this subcommittee hearing, and to you and Senator Lummis for spending so much time to try to bring focus to the needs for authorizations. And thank you to the witnesses.

I think I will start with you, Mr. Bridenstine. Obviously, one of the past times you came before this committee, you spoke very fervently about the need for authorization, and yet it has been since 2017 since we have actually had an authorizing bill. So, I sometimes feel like NASA wants to have the money without the authorization. That is, that it works, not to resolve the conflicts that we have with members that, basically, it just realizes, as long as you can just get the dollars it is OK. Well, that is not OK. And so, I just want to be clear with the panelists, if you could give me some feedback on some of these issues. Do you think that we need an authorization bill?

Mr. BRIDENSTINE. Yes, 100 percent.

Dr. DITTMAR. Yes.

Mr. GOLD. An authorization bill is vital to send messages to our partners and rival nations, with unity purpose of Congress, yes.

The CHAIR. Dr. Sanders?

Dr. SANDERS. Yes.

The CHAIR. OK. So, do we need to have more testing and analysis of rocket launching and the capabilities that we are looking for in the next system?

Mr. BRIDENSTINE. Do we need more testing and analysis?

The CHAIR. Yes.

Mr. BRIDENSTINE. You can always have more testing. There is not doubt. And more analysis. At some point—and I like the way Patricia Sanders mentioned it earlier—we have to manage risk. So, it depends on what system we are talking about and—but certainly, there is value to more testing.

The CHAIR. Dr. Dittmar?

Dr. DITTMAR. So, having been involved in the development of the space station, I would just echo what Jim said, which was——

The CHAIR. I am just talking about—I will get to space stations in a minute.

Dr. DITTMAR. Specifically rockets?

The CHAIR. Yes.

Dr. DITTMAR. No, what I was going to do was simply speak to engineering. OK, so—so it is always useful to have test data, but there is also an appropriate use of test data and then, a point at which just more and more testing simply adds cost and schedule.

The CHAIR. And Mr. Gold?

Mr. GOLD. Senator, this country has not had a human space flight program beyond low Earth orbit in a very, very long time. We are going to have to relearn some critical lessons, as well as developing new technologies and new systems along the way. So, testing and making sure that we have a safe system that is robust and effectively competes with China and other nations is going to be critical, leveraging the wonderful private sector companies in your State, as well as those three that Senator Hickenlooper mentioned, of Redwire in Colorado.

The CHAIR. OK. Dr. Sanders?

Dr. SANDERS. Strong proponent for adequate testing and analysis. As Mr. Bridenstine had said, it is not possible ever to completely and fully, totally eliminate risk and you want to manage it. But the way to manage it is through knowledge and we—as much knowledge as we think we have about systems, we continue to learn things about parachutes, about compatibility of the components, materials——

The CHAIR. Do you think NASA Houston needs to have more oversight over the Artemis program and its schedules and launches?

Mr. BRIDENSTINE. I think their——

The CHAIR. Than is currently provided.

Mr. BRIDENSTINE. I think their—well, oversight from the Congress or——

The CHAIR. No, from NASA Houston.

Mr. BRIDENSTINE. Oh, from NASA Houston, for the Artemis program.

The CHAIR. Yes.

Mr. BRIDENSTINE. Well, certainly there are a lot of different centers that are involved in the development of different components of the Artemis program. I think it is important that NASA has a robust capability to do the integration and I think, until recently, that has been lacking. But my understanding is—and I am not there anymore, but my understanding is they have really plussed up the systems integration piece that has been missing for a while.

The CHAIR. Dr. Dittmar?

Dr. DITTMAR. With apologies, Senator, I am not as—I am not as current with what management has distributed across the center.

The CHAIR. OK. Basically, I am bringing up topics that are part of the dispute between what the Senate and the House wants to do on a NASA authorization. So, I am just trying to get your viewpoints on that because part of the issue is, we have to debunk, you know, where we are. We cannot do an adequate oversight job if we do not have an authorization bill. We cannot come back on the measurements if we do not have some input here about what kind of structure we need.

But we are sitting here with, obviously, a new—as Mr. Gold was saying, a pretty big, new adventure, at least from the number of stops we are talking about. And then, what we are talking about going beyond the moon. To say nothing of we really do not have any accurate dates or cost estimates or what we want out of each of these systems. What do we want out of the launch systems and the capabilities? What do we want out of the ISS and its capabilities? What do we want on the lander system? What do we want, you know, on the beyond lander—I am sorry, beyond the Moon?

So, here we are having this discussion without reference and without oversight, really, because we do not have an authorization bill that is on a piece of paper, who is in charge, who is going to answer these questions, and how are we going to have this debate in Congress? So, what is coming across is the amount of money that people want to keep going, but then, it is always not enough. And then, the choices that people make do not necessarily adhere to redundancy and resiliency.

So, I think it is just imperative that we get an authorization bill. And so, I am trying to figure out from you all, what you think some of these stumbling blocks really, truly are between our colleagues. So, Mr. Bridenstine?

Mr. BRIDENSTINE. Ma'am, I think—I think that is all—I agree with everything you just said. I think that is exactly right. And I would also say that one of the biggest values of having an authorization bill—yes, oversight is a big piece of it, but for NASA to have continuity of purpose over time requires an apolitical, bipartisan consensus on how we are moving forward. And if we can put that in an authorization bill, it sends a signal to everybody, globally, that we have resolve to accomplish these objectives. And then, as this hearing is titled, we have the opportunity to go get international partners.

If we do not have that resolve—if the international partners do not trust that we are actually going to accomplish what we are saying we are going to accomplish, the first thing they are going to do is, they are going to go to China, which already has, you know, what they call the Chinese International Space Station and now they have entered into an agreement with Russia for going to the moon and establishing a lunar base. Our international partners have a history of partnering with other nations when we do not have that continuity. So, the authorization is important for that purpose, too.

The CHAIR. OK. Mr. Gold?

Mr. GOLD. Senator Cantwell, when Administrator Bridenstine sent me overseas to negotiate the Gateway Commitments, it was because of statements that you made that we were able to complete that negotiation. I was told outright, by an international partner, "Why should I believe anything that you are saying?" Relative to Gateway, Artemis, and NASA's plans. And it was only by arguing the bipartisan support that Artemis had, that we were able to bring them and their collaboration to the table and prevent them from going over to China. Without an authorization bill, those doubts will continue to fester.

The CHAIR. Well, one of the reasons why NASA's authorization is on the USICA, America's competitiveness bill, is because we have passed this twice now, out of the Senate, to have no results in the House. So, we are very adamant here, in a bipartisan fashion, what needs to be done and I am just trying to use today as a way to figure out what is this stalemate that we have with our House colleagues truly about.

Now, I do not know if anybody wants to talk about the nature of a public lander—that issue. But obviously, that is one of the stumbling blocks, as well, is that people would like to have a process and go back and look at a public lander. Although, I think some people are saying, you know, the IP would belong to the government, all sorts of—so, how do we—how do we get a resolution of this issue?

Do we have other members waiting? Sorry.

Senator HICKENLOOPER. I think you are on such an important point that——

The CHAIR. OK. OK, thank you. I do not know if we have colleagues here, so I do not want to hold up somebody if they are on——

Senator HICKENLOOPER. No, you are—there is no one in the queue.

The CHAIR. Thank you. That is——

Senator HICKENLOOPER. They are all voting. They are voting.

The CHAIR. — too bad for—too bad for the witnesses. Yes, go ahead, Mr. Bridenstine.

Mr. BRIDENSTINE. So, when we think about a human landing system for the moon, I think it is—we need to look back at the Commercial Resupply Program and the Commercial Crew Program. Both of those have proven to be very resilient. Even when one partner had a challenge, the other one would step forward and continue to move forward. We saw that—you go back to 2014 on commercial resupply of the International Space Station, we saw, you know, at the time it was orbital ATK. They had a problem resupplying the International Space Station in rocket failure. A lot of cargo blew up. And then, in 2015 SpaceX had a rocket blow up with the international docking adaptor and other cargo.

But we had a resilient program where we could—we had dissimilar redundancy and we were able to use an Atlas rocket with a Cygnus cargo capability and basically, resupply the International Space Station. And we had international partners on that, Japan, that helped, you know, support the resupply and even Russia supported the resupply.

So, the answer is we need this similar redundancy. Having a single provider for a human landing system, I think, imposes risk. That risk is budgetary. That risk is schedule. That risk is oversight. That risk is transparency. And I think, if we have competition, multiple providers that are competing on cost, on innovation, on safety, and those providers operating commercially are getting customers that are not necessarily the government, I think that is a good thing for the program. I think that was the original intent. I think that is why when, you know, Congress initially funded the Artemis program there was an anticipation there would be two, at least two, in the—in the competition. I was fully supportive of that and that is what we were pushing for.

The CHAIR. I am more trying to get—our colleague, the Chair of the House committee wants a public lander. And so, I was trying to get comments about that.

Mr. BRIDENSTINE. I think—I am going to say things I have said before and it is going to, maybe, get me in trouble with members of the House. I do not know. But I will tell you that I think any time you introduce a competition between the government and the private sector, the government wins, and the private sector loses. I think that is a challenge. I think—I think we have enough commercial capability now to have two commercial human landing systems. And I think, if we had adequate funding—you know, whether it is coming from the Executive Branch that is not providing it, or Congress, I think the adequate funding for two commercial human landing systems, at least, maybe even more, would be the right solution.

Mr. GOLD. Senator, I think the private sector has proven itself. It has delivered cargo to the International Space Station. It has now delivered crew to the International Space Station. The private sector can and will be able to meet that mission. What we are missing now, as you mentioned, dissimilar redundancy and direction and funds. And those are two things that we certainly hope Congress will be providing via an authorization bill. And certainly, we are advocating for the House to move and move quickly.

The CHAIR. Well, we are trying to resolve these issues. So, the point you should take away—and NASA, really, if they are listening should take away is, we need an authorization bill. We are not going to continue to have this game where you just get appropriations. It is not going to serve the Artemis program or NASA well, but we have to resolve this issue. And frankly, you know, it is frustrating to me to see the spirit of NASA turned into, you know, the cheering of billionaires or competition with each other, because I am pretty sure that is not what our mission of Artemis is about. And so, somehow, we have gotten away from our focus, you know, on what we are trying to do from a technology perspective. And I believe in NASA's innovation in technology. I want them to apply the same spirit they had in fixing the problems on the Apollo project, to fixing and getting this authorization done. But if we have to address this public issue and get our House colleagues in a room and figure out what is the sticking point here, we should do better than just—just dismissing it and saying the private sector can do better. We just had this debate with the FAA and oversight of the aviation sector. We clearly saw where there were pitfalls.

So, we have to figure this out and resolve our differences so we can get legislation, so Congress can do its oversight job, so that Senator Hickenlooper can continue to play a great role here in pushing this, and then, we can get to the international aspect of this. But right now, we are—you know, we are—it has been since 2017 since we have had a bill. And so, you ask yourselves why— why are we in this situation where the dates that are on a piece of paper right now, make no sense, as it relates to the Artemis program. Everybody knows that. So, how is that a good strategy? So, let us resolve these issues.

Dr. Sanders, do you have any suggestions about how to resolve this issue about a public lander versus commercial landers? Do you have any input? Or Dr. Dittmar?

Dr. SANDERS. I think that, yes, the commercial industry is capable—increasingly capable. But NASA always still has to be accountable for what—for the success and safety of their mission. So, it is important for them to be able to have enough influence and interaction in order to have that—ensure that accountability.

And also, NASA has a great deal of experience, a great deal of talent from—that they have acquired over decades. And so, there is a time for collaboration as this is going on and we saw that in commercial crew. There were times when SpaceX and there are times when Boeing have had—had problems to solve and NASA has been able to help them solve those because of the experience they have.

So, I do not think it is a clear, straight, turn it over to commercial, nor is it a clear, straight, that NASA develop everything by itself.

The CHAIR. Well, that is why I predicated my question earlier on that list and wanted to know what you thought about giving NASA Houston a larger role on that oversight. Look, we have seen where the same mistake was made by the FAA in deferring too much to aviation manufacturers, when it came to the oversight. So—and we have huge technological advances. And so, there is a lot to know and to be tested.

So, we need to get—have this structure—I want a very strong NASA oversight of these companies—very strong. I am not supportive of NASA stepping away and turning it over to the commercial side. But we have to figure out what that looks like, from an authorizing perspective, and put that in a piece of paper.

Mr. Gold or Dr. Dittmar?

Dr. DITTMAR. I would just add quickly, with regard to the lander, I do not know what the particular rationale is underlying the various positions, but one thing that might be pointed out is, if NASA is engendering—and I am in agreement with Jim, I think we need at least—two would be good. We definitely need redundant capability if we are going to take this approach. If—you need to do a path, basically. So, if you are going to do that—you know, one thing the government can always do is assert, what amounts to, imminent domain, right at first use, so that if—and I am speculating, if the issue having to do with the public lander has to do with whether or not the government could count on being able to use it, as it saw fit, in the same way that it can with regard to a government loaned asset.

And there has been a great deal of discussion. I am one of the people having the discussion over the last several years, having to do with what is the appropriate role of government in signaling, particular to international allies, as Mike has already pointed out, and adversaries.

What is the full faith and measure of the U.S. Congress mean and what is the difference between how it is that you see a government owned asset versus a public asset, OK, versus how you see a privately owned asset?

Maybe one way to have the discussion is to talk about what rights the government has to assert in the case where it needs to use, OK, and to, in fact, basically say to private companies, "Sorry, we understand you have other customers. We understand you have business and agendas. But we have to set those aside, under these circumstances." I do not know if that is a path forward, but it is just an idea.

The CHAIR. Well, I think oversight is the question here. My sense is there is a feeling of loss of oversight with these commercializations. And, as I said, we have dealt very deeply with this, as it relates to the FAA and to the manufacturers. And you have a lot of people, even within the organization at the FAA, stepping away saying, "Oh, they know better. Let them go ahead."

Dr. DITTMAR. Yes.

The CHAIR. And we need a very strong NASA and very strong NASA oversight.

Dr. DITTMAR. I believe to do that, what you also need to have, and when you think of authorization, is very clear statement of objectives. What are the objectives, OK, and so far as how the government sees them and how Congress sees them? Because without that, it is very difficult to even begin to be able to do oversight because you do not know what objectives you are trying to meet.

The CHAIR. You could not have said it better, thank you.

Mr. GOLD. And Senator, perhaps you have the solution already, in terms of the compromise with the House, relative to oversight. I believe the question of government versus commercial is a false dichotomy, that we are stronger when we are working together. As Mary Lynne mentioned—Dr. Dittmar, that NASA has got so much incredible experience and the private sector has innovation, dollars, affordability. We need to combine that effectively and, hopefully, going to the House, bolstering oversight and insight of the programs and having two entities moving forward, I hope, could address the issues that you are raising.

The CHAIR. You are raising an interesting point, but I think no one's against companies going out there and doing commercial space travel. OK, go for it. But we are talking about how we are now going to conduct our next Artemis mission.

Mr. Chairman, you have been so lenient, and I see my colleague has returned. So, I am sure there are more questions by my other colleagues. Thank you so much to the witnesses and thank you for your diligence on trying to get this authorization over the goal line. Thank you.

Senator HICKENLOOPER. Thank you, Madam Chair, and I was copiously taking notes, as I trust you all were with her questions. I also have to vote, so I am going to turn this—the chairing of this meeting over to Senator Lummis, our Ranking Member, and then, she will proceed. But I will be back. So, you know, do not think you are going anywhere soon.

[Laughter.]

Senator LUMMIS. Well, thank you, Mr. Chairman. And thank you again, panelists, for tolerating our vote series that is interrupting the conversation. Former Congressman Bridenstine remembers those days very well.

First question is for Mr. Gold, and it has to do with the importance of strong public-private partnerships and a good line of communication between NASA and the commercial space industry, particularly on regulatory and policy issues. So, Mr. Gold, I would like you to talk a little bit about the importance of the NASA Advisory Council's Regulatory Policy Committee.

Mr. GOLD. Thank you, Senator. Public-private partnerships in America are the envy of the world. It is incredible. This really, second Golden Age that we are having in space due to, not just the private sector, but the support that NASA has given—that combination of capabilities. Elon Musk has gone on the record that there would not be a SpaceX if it was not for NASA. So, being able to combine public and private sector benefits and advantages are so critical.

And one of the other aspects that separates us out from foreign competition, other nations, are FACAS, the Federal Advisory Committees, that we have an explicit and official ability to combine private sector feedback and advice. And in this world, as Senator Cantwell just mentioned, where we are dealing with some pretty extraordinarily policy issues, the NAC Regulatory and Policy Committee is really the only place where NASA can go to get private sector feedback that represents, not just companies that, you know, are new—like, SpaceX and Blue Origin, Virgin Galactic—but also, Boeing and Lockheed, the traditional and terrific space companies that have brought us so far.

So, having that common ground, where both the new and the traditional space companies can come together to advise NASA is just critical and will allow us to leverage the power of the private sector, in support of NASA's goals.

Senator LUMMIS. Thank you. Would anyone else like to comment on that? Well, thanks. That is a pretty strong statement in favor of their work. So, thank you.

Mr. GOLD. The embarrassing thing to me, Senator, is that two of the members of that committee are now astronauts that we had. So, I am falling behind my colleagues who are on that committee——

[Laughter.]

Mr. GOLD. Audrey Powers, Sirisha Bandla.

Senator LUMMIS. No worries. You bring your own qualities to that. This is for anyone who wishes to jump in on it. How would you say China and Russia manage their orbital debris? And are there things we can do to encourage them to do more?

Mr. BRIDENSTINE. So, everybody claims that they follow the, you know, the guidelines that NASA creates and is adopted broadly by the U.S. Government. And then, it goes and goes to international fora and people say, you know, we all agree on these guidelines. In practice, do all nations follow those guidelines? Absolutely not. And so, that is a challenge. It is one of the reasons, when I was a NASA administrator, I got Mike Gold engaged in the Artemis Accords. We needed to use the Artemis program—and I think this is an important thing. We—we look at the Artemis program, we get all these countries that want to be part of the Artemis program. How do we leverage that to create an international environment that is conducive for the future of human space flight? And so, we put in there that you had to adhere to the debris mitigation guidelines that are set forth by NASA. And interestingly, everybody has already agreed to that, so nobody could ultimately disagree. But here is the thing. If you want to participate in Artemis, now you are committed. You have got to follow these guidelines. And I think—so, the question is, how do you—how do you compel other nations to follow what the guidelines are?

And, oh by the way, and this is another challenge that we have, as I mentioned in my opening statement, in this country we have, you know, these mega constellations now, that are also placing at risk, low Earth orbit. And I am not naming anyone. There are lot of them in this country. And as I mentioned earlier, there are a lot of other countries that have constellations, as well. Rwanda, as I mentioned, 327,000 satellites being filed for at the ITU. So, we do not have the authority, as a Nation, to turn around and tell other countries, "You are not allowed to have your own constellation," when we ourselves are licensing exactly the same things.

Senator LUMMIS. Yes.

Mr. BRIDENSTINE. So, the solution set, Ranking Member Lummis, is to recognize that there is a limitation to how much stuff can be put into any orbital shell. There is a limitation there. Once that— and, by the way, I do not know that that has been recognized by anybody at this point. But once it is recognized, we need to define what those limitations are. And those limitations can be determined by the cross-section or the size of the satellite, the mass of the satellite.

So, we think about how much stuff can you put in any orbital regime and then, we say, OK, now that we know that there is a limitation, we have defined what that limitation is, there needs to be a process by which we allocate access. When I say we, there needs to be an international kind of effort to this, where access is allocated in an orderly way. And when the United States gets its allocation that we, in fact, do allocate in a market competitive way, where we allow access to a number of different companies that are trying to accomplish the same thing, so we do not end up with monopolistic behavior.

I really think, if you look at the—kind of, the way the ITU works for geostationary orbit, I think that that might be a good model for low Earth orbit. It is going to have to be because the other result is this. We are all launching stuff into space, and it is going to— it is going to be not good. It is a race.

Senator LUMMIS. Yes.

Mr. BRIDENSTINE. And when you have a race like this, and everybody is operating—it is the prisoner's dilemma. Everybody is operating to benefit themselves and, at the end, everybody loses.

Senator LUMMIS. Mm-hmm.

Mr. BRIDENSTINE. It is the tragedy of the commons and I think that is what we are facing right now in low Earth orbit. And there is no regulatory regime that manages it appropriately, and there is certainly not an international regime to manage it appropriately.

Senator LUMMIS. OK, so the ITU——

Mr. BRIDENSTINE. International Telecommunications Union. It is a subset of the U.N.

Senator LUMMIS. And would the U.N. be the proper organization to also be the umbrella organization for this?

Mr. BRIDENSTINE. Well, so wow. [Laughter.]

Mr. BRIDENSTINE. It is a challenge, for sure, because I would say that the ITU works, and I will tell you why. And Mike and I were just talking about this earlier. There are countries that are at war with each other, that want to have a satellite in geostationary orbit, and they go to the ITU to work it out. So, it works, and it has worked very well in geostationary orbit. Right now, there is nothing for low Earth orbit and I think that is a problem.

Senator LUMMIS. Does anybody else want to weigh in on this before we move on?

Mr. GOLD. Senator, if I may. Space should be about joy and discovery, not danger and fear.

Senator LUMMIS. Mm-hmm.

Mr. GOLD. And per your question about China and Russia, that is what we are getting with their debris. You know, now I am going to steal from the former Administrator there, who would often talk—and we would be remiss if we did not mention ASAT testing in the context of this discussion. That due to a Chinese ASAT test, there is debris that is still up there. And I believe it was America that had to warn China that there was debris, created by their own ASAT test that was threatening their own space station. You know, the Foreign Minister would always point that out.

So, they are not particularly responsible actors. There is Chinese debris, you know, debris that comes down from every launch. We do not know where it is going to land, often. And last couple launches, it could have even hit America. We do not know.

So, as Jim was describing, what we need to do—and I do not think the U.N. is the entity to do it in, at least initially. We need to do what we have done with the Artemis Accords, and I am still grateful for the Administrator—former Administrator—for that opportunity, is to build a coalition of the willing, in terms of what good looks like. Lead by example on the debris issue and then, use that to force United Nations, or form a new kind of ITU, to be able to develop it.

Also, debris represents not just the problem, but as a private sector Redwire guy, it is an opportunity to innovate, to develop new technologies that can address this issue. We should have active debris removal missions, so that we can develop new systems that not only will help address the debris problems, but will create new sat ellite technologies and other systems that can benefit a wide variety of economic national security areas for the country.

Senator LUMMIS. Mr. Bridenstine.

Mr. BRIDENSTINE. First of all, I want to double down on what he said. I think Mike is exactly right—Mr. Gold, as his nametag says, is exactly right.

When we talk about—think about the FAA and we go back to the 1950s when the FAA was created. In those days, the United States did it alone and we were making great advances and it worked. We were able to have a lot more airplanes into a lot less space. Interestingly, because we took the lead, and we took that model around the world, an organization formed called ICAO, International Civil Aviation Organization. And they took the leadership of the FAA and they applied it globally. So now, when I fly my airplane, Navy pilot by trade, when I fly my airplane all over the world, I know exactly how to fly in whatever country I go to because we are all following the same rules. But it would not have happened if we would have, you know, gone to, you know, some international organization and said, "Here is what we need to do," and try to get agreement from our competitors. That would not have worked. But when the United States takes the lead, ultimately there becomes a standard that other countries can join. And I think that is exactly the idea that Mike was talking about just a few seconds ago.

We have to lead on this and when we do, I think we are absolutely going to, you know, change the world so that all of those constellations, that we all believe in, for low latency, high throughput communications, can be successful.

Senator LUMMIS. Well, this is a fascinating subject and I hope to pursue it further. But I want to recognize my colleague. This is Old Home Week. Not only is Mr. Bridenstine a former Congressman, but of course, so is the gentleman from New Mexico, Senator Luja´n, who we recognize now.

Statement of Hon. Ben Ray Luja´ N, U.S. Senator from New Mexico

Senator LUJA´N. Thank you to our Ranking Member and to our Chair for this important hearing. Holding this hearing matters to us not just here in America, but in New Mexico.

Now, in New Mexico, we have an incredible innovation ecosystem that supports the space industry. There is substantial research being done at the universities, the Air Force research labs, the Satellite Space Operations office, and two of our three NNSA Department of Energy National Laboratories. We have the home to the first commercial space port in the United States, where Virgin Galactic successfully launched its most recent space flight, as well.

Now, to our former Administrator and colleague in the House, small launch companies are facing pressure as governments all over the world subsidize their launch industries. To ensure that NASA keeps a capability that many, including yourself, have said is critical, do you think NASA should expand the scope of the Rideshare office to be a small satellite launch office, charged with assessing mission needs relative to all available launch options, in order to determine the best solution on a mission-by-mission basis?

Mr. BRIDENSTINE. So, I will start by saying this. When I was in the House, I was a big champion of the Venture Class Launch Services Program, which was basically a bill that enabled NASA to—to support small launch companies in a, you know when you are trying to launch a small vehicle to a very specific orbit, we are talking 500 kilograms to 1500 kilograms, we did not have the capability, you know, five or 10 years ago to do that to a very specific orbit. So, the Venture Class Launch Services Program was used to let the private sector know that the U.S. Government has an interest in seeing the ability to launch small payloads into specific orbits. And it worked and now, we are seeing a lot of these companies develop that capability.

I would also say, when I was the NASA Administrator—you mentioned Virgin Galactic. I mean, we had value from the fact that we could put our payloads at NASA on those vehicles. Of course, it benefited them, but it benefited NASA.

The challenge that we had at NASA is, you know, you can get microgravity testing with drop tests, where you might get a couple of seconds of testing. You can do parabolic trajectories, where you might get, you know, minutes, like, eight or 10 minutes—or, I am sorry, 30 seconds, or whatever. But then, when you talk about, after that, you have got to go to, you know, low Earth orbit. So, a lot of these small launches, whether they are usually suborbital, there is opportunity there that, I think, you know, Virgin Galactic, Blue Origin, are making available to NASA.

So, I do believe the United States of America has an interest in making sure that these companies are successful, because it is in the interest of our country that they are successful, for a whole host of different testing and science reasons. That being said, as far as that particular program, I would love to give you an answer— maybe do a little more looking into it and give you a more specific answer about that specific question.

Senator LUJA´N. I appreciate that. And the next question I have is for both our former Administrator, Dr. Dittmar, and Mr. Gold. One of the bills that I am working on is building off an idea that has proven to be successful with other Federal agencies, and that is establishing a foundation. We are working on one with the Department of Energy. One exists with the National Institutes of Health. And it has been able to strengthen public-private partnerships to be able to attract that private capital and to make these investments.

Now, as the former Administrator and leaders in the private industry for space, do you agree that if we fail to harness the full commercialization potential of NASA, we will fall behind countries willing to do so?

Mr. BRIDENSTINE. I absolutely think so. And I think, you know, my friend, Mike Gold, talks about it as, you know, our competitors in the world are never going to out entrepreneur the United States of America. And the best innovation that we see coming usually comes from small, innovative companies. And so, I think, in many cases, it is in our interest to make sure that we have those public- private partnerships for these unique opportunities.

It is important to also remember that—that when we have those public-private partnerships, the goal is for those companies to go get customers that are not the U.S. Government. And if they do that, it drives down the cost to the American government. And we need to make sure, when we enter into public-private partnerships, that there is still competition in the marketplace, so that the competitor is competing in cost and innovation and safety.

Senator LUJA´N. Dr. Dittmar?

Dr. DITTMAR. And so, I would just agree with everything that was just said here. And I think public-private partnerships—you know, public-private partnerships, they have become, you know, quite the thing to talk about in space over the last 20 years. But they have been extant in this country since we started the country. They have been used in a variety of different ways to create infrastructure.

And we are looking for a whole ecosystem in space, right? One that supports everything from ground test equipment, all the way up to orbital space flight and then, beyond. There are opportunities we probably have not even thought of yet, in terms of how to utilize public-private partnerships to advance that.

What the United States needs it to be able to have enough strategic flexibility that it can pull whatever levers it needs to to, sort of, be able to advance the private sector, as well as advance government—government intention and have those things work together. So, yes.

Senator LUJA´N. Appreciate that. And Mr. Gold, I am going to ask for you and Dr. Sanders to maybe submit your answers to the record on that particular question. And then, the other question I have for the panel, that I would ask for you to submit to the record—we will submit it, is the importance of spaceports and especially the example that I shared with Spaceport New Mexico about the continued improvements to America's spaceports are needed to grow, not only our industry partnerships, but also for safe and sustainable government use.

So, I will submit that to the record.

I want to be respectful to Mr. Blumenthal's time, as well.

And I want to thank the Ranking Member for her acknowledgment today.

Senator LUMMIS [presiding]. The Chair recognizes Mr. Blumenthal.

Statement of Hon. Richard Blumenthal, U.S. Senator from Connecticut

Senator BLUMENTHAL. Thank you, Madam Chair. Nobody is going to out entrepreneur the United States sounds like famous last words. And the idea of a public-private partnership, when it comes to space, the idea of an ecosystem brings back memories of my law school days when people were saying we should have a course at Yale Law School on the Law of Space. Only at Yale Law School would anyone make that suggestion. And it was dismissed as one of those bubblehead, imaginary topics.

But there is no real law of space, is there? I mean, and should there be? Should we be thinking about, if we are going to have commercialization in space? You compared it, Administrator, to flying into other countries. But there is a means of enforcement there. You can ground pilots. You can take action that provides enforcement of rules of the road, so to speak—rules of the sky, rules of landing and taking off, and equipment, and so forth. But there is not any in space. And so, I mean, if there are three or four commercial flights going up on the same day, maybe from the same place, you have got a lot—you are going to have a lot of objects up there. You already have a lot of objects bumping into each other, except they just do not have human beings on them. And having human beings, and having commercialization for expanded purposes, either of surveillance or other missions, complicates it. Should we be thinking about laws and enforcement mechanisms in space?

Mr. GOLD. So, Senator, as the only recovering attorney on this panel. I may go ahead and take that one. And for the record, it is not just Yale. I am actually giving a lecture at Georgetown Law School tomorrow on Space Law. So, it has certainly taken—taken root.

I do want to open with that there are laws in space. There is the Outer Space Treaty of 1967. There is the Registration Convention, the Agreement on the Rescue of Astronauts. So, there are a series of international agreements that set rules that nearly every country, or spacefaring country, certainly have agreed to abide by. However, there is a substantial gap that you are pointing out now, where, in the Outer Space Treaty, in Article 6, it requires authorization and continuing supervision of private sector activities.

And we here in the U.S. have yet to define how that continuing supervision will occur for non-traditional activities—be it satellite service, which we have mentioned, private sector space station, orbital debris removal, rovers—commercial rovers on the moon. We have not defined that yet. We have been using Band-aid solutions, via the FAAST and the FCC. But one of the most important things that Congress could do would address this issue, just to create a process that is predictable, and we know what to do here in the U.S. Because, frankly, it is a competitive issue. That predictability is key for investment and if we want to keep our country's here and keep innovation and keep entrepreneuring here in America, we need a predictable, reliable, and transparent regulatory system for commercial activities, which we do not have explicitly, yet.

Senator BLUMENTHAL. I think that is an excellent answer and obviously, I should have been more definitive or more specific in saying there is no law. There are, in fact, gaps in the law. Which means I probably should attend your lecture tomorrow at Georgetown. And I realize that law is the least glamorous or exciting aspect of space exploration. But I am very interested in your answer that, as in many other areas of life, certainty and predictability are very important for the risk takers and the scientists to make progress.

Mr. GOLD. Exactly. If you are going to invest, the last thing you want to see is an unpredictable regulatory environment. If you are going to ensure an activity, you need that certainty. So, I know it is not as sexy as fun, but it can be as important as any technological development to space not only occurring, but particularly occurring here in the United States.

Senator BLUMENTHAL. Spoken like a true recovering lawyer. Thank you. Thank you all for your testimony today and it has been very enlightening. I have been following it from afar, in between hearings, and we really thank you for being here.

Senator LUMMIS. The gentleman yields back and the gavel is returned to the Chairman, the gentleman from Colorado.

Senator HICKENLOOPER. Thank you, Ranking Member Lummis. What did I miss? No.

[Laughter.]

Senator HICKENLOOPER [presiding]. We have a couple of people that are trying to get in remotely, and luckily, I happen to have more questions and I think Senator Lummis might have some—a couple more questions, too.

So, if I can re-enter the low Earth orbit of information, and let us go back to, Mr. Bridenstine. You spoke, I think, compellingly about the overcrowding in LEO and you did not even really get into that whole notion of, once you have collisions, how many more fragments are created and that you end up with a cavalcade of collisions. And that 1.5 percent really gets scary because that is too high, in and of itself, and I followed your math of how we get to 1 in a million. But there are a lot of probabilities between here and there and I think that is—it is unacceptable, when you begin to look at what—how rapidly you could escalate.

So, anyway, I think—again, how would you look at what should Congress be doing? Who should be—have the regulatory responsibility there? And we are all—we are fully aware of the discussions there, but I would love to hear your opinion.

Mr. BRIDENSTINE. So, the—right now the regulatory authority for debris mitigation falls to the FCC, the Federal Communication Commission, and that is based on a law passed in the 1930s. Mr. Gold could probably tell me the specific date. But there was a law passed in the 1930s that gave the FCC the authority to regulate Spectra. And from that, you know, satellites, you know, emit, you know, all kinds of transmissions. And so, from that, they say we have the ability to regulate satellite, you know, debris, and orbits, and that kind of thing. Which, by the way, it is good that they are doing it because if they were not doing it, nobody would be doing it. So, they extrapolated the law to take that authority among themselves.

I do not know—in fact I am confident that the FCC is not the agency that should be doing that. And the House and Senate have passed a bill that was signed that gives that authority to the Commerce Department. That has not been adequately funded at this point. They need, you know, resources to stand up a team of folks that can actually put into place the regulatory environment for this—for this capability.

Senator HICKENLOOPER. So, are you agreeing that that notion of having a separate escalating one of the departments from Commerce up to—

Mr. BRIDENSTINE. Yes. I know exactly what you are saying. So, the Office of Space Commerce.

Senator HICKENLOOPER. Yes.

Mr. BRIDENSTINE. Yes, so 100 percent.

Senator HICKENLOOPER. Office that is the word I was looking for, office.

Mr. BRIDENSTINE. Yes. The Office of Space Commerce is, right now, under NOAA. It is not—and look, we all love NOAA, but it is not their highest priority to do space situational awareness and space traffic management, let—and debris mitigation and remediation, as you mentioned earlier. So, since that is not—NOAA is not, probably, tremendously enthusiastic about that.

I do believe it is a high enough priority. It needs to report to the Office of Space Commerce. It needs to report directly to the Secretary of Commerce and be responsible to the Secretary for achieving the outcomes that we are looking for.

As far as what the FCC needs to do right now, given the risk that is in front of us, they need to consider that even satellites that can maneuver do have a collision risk. And that risk does not just come from other satellites, it comes from lethal, non-trackable debris. For every piece of debris, we can track that is 10 centimeters and bigger, there are, you know, 10 to 100 pieces of debris that we cannot track. So, they need to take that into account when they start thinking about the collision risk for—for, you know, low Earth orbit.

Once we recognize that there is a lot of risk here that, right now, is not being accounted for, I think it is important for us to say, OK, how many objects are we able to put into any orbital shell? And once we define that, then there needs—this is where the international pieces comes in. We recognize that there is a limitation. The question is, how do we allocate access to those orbital shells? And we need to have a process by which nations can get access to that and the United States of America needs to make sure that its allocation is, of course, adequately spread among, you know, competitive companies for a competitive marketplace.

Right now, none of that is happening. It is a race, and the race is, ultimately, going to result in disastrous consequences, if we are not careful.

Senator HICKENLOOPER. It reminds me a little bit of the old—do you remember the—as popularized in movies, via Hollywood, the Oklahoma Gold Rush—Land Rush, where they would all ride out and they had to stake off their little territory and they would be shooting each other and—I mean, it is, to a certain amount, out of control.

Dr. Dittmar, just continuing on that, this importance of having international agreement on how we are going to regulate this debris that is up there already. And this is one of those things—I mean, the first thing, when I have nightmare, it is usually about cybersecurity. But this is now becoming—it is—it is equally imposing onto my sleep patterns.

So, how would you look at the—what should Congress' role be toward building, or moving toward an international agreement on the appropriate regulation of space debris?

Dr. DITTMAR. So, following on conversations that we have, sort of, already been having here on the table, one of the things that is really critical about authorization acts is that they are signaling devices. They signal not just to the United States, with the intent of Congresses and especially its investment in continuity of purpose, but they also signal to the international community. So, othercountries look to see what Congress does. They look to see what is written into policy that is then expressed in an authorization act. And one thing I would like to see—I am in complete agreement with Jim, with regard to elevating the Office of Space Policy, completely making sure that it is consistently staffed across administrations, to the extent that you can. You can certainly express an intent, OK, and the intent of Congress, that it is adequately staffed and resourced with the appropriate expertise, OK? And then it is a direct report to the Department of Congress and the head of the Department of Congress. And so, doing so, while you are investing it, OK, with the authority to begin, then developing the sorts of regimen.

You know this has been—Jim has talked about one—you know, one approach, which is sort of to look at, literally, what you can get into orbital shells and start thinking about regulation that way. But to invest them with the authority to be able to actually begin to be able to work in that way. Just doing that, OK, will convey to the United States how serious, I mean, how seriously the United States is taking this.

It is going to also help. There is a big PR war going on. I mean, it is a competitive issue. But there is a huge PR going on—war going on. You just, basically, need to send a message to essentially all the stakeholders here. "All right, look, no, we are serious about this. We are going to start pulling this together. We need to develop a regime. We are going to invest this office with the authority to be able to do that." And it also signals, to the rest of the world, now, the United States is serious. We are going to take a leadership position here. Here is how is going to be, who is going to be accountable for actually being able to do it? And I know that sometimes, from the outside, that does not seem like much, but it really is a big deal.

Senator HICKENLOOPER. Oh, I buy that. And just as a frame of reference, the Office of Space Commerce, I believe, still does not have an executive director. So, we are in the market of hiring. So, if you have networks or connections, step forward and take one, you know, take one for the country, if you could.

All right, one more question and I will turn it back to Ranking Member Lummis. And I guess I will ask Mr. Gold this, since we are looking at global leadership and technological leadership. You know, obviously leading-edge technologies are evolving and innovating at lightning speed in our commercial space industries. Which technologies do you think are most critical to make sure that the United States does not cede the leadership role to our competitors, such as China?

Mr. GOLD. Yes. Thank for such a terrific question, Senator. There are really two areas that I would highlight. One is, we are at a period where robotics and satellites are merging to create something entirely different. At Redwire, for example, I mentioned the Archinaut satellite, which is—uses robotic arms to build itself. We are also developing satellites that will literally print themselves, leveraging the 3D printing technology that we have done on the International Space Station. So, when it comes to this orbital servicing, assembly, manufacturing, we cannot fall behind China andothers in this area. And the key to developing that is more public- private partnerships.

The other area I would cite is microgravity. That the micro-gravity environment is a whole new arena for scientific and commercial research. I believe it is why China is investing in their LEO space station. And we are just beginning to understand the biotechnology, the fiber communications development, new super- computer chips. Everything functions differently in low Earth orbit and the microgravity environment. And we must invest and develop those capabilities because it is going to change everything.

So, those are the two areas I would recommend focusing on.

Senator HICKENLOOPER. All right. They are good ones. I will turn it over to Ranking Member Lummis.

Senator LUMMIS. Thank you.

Senator HICKENLOOPER. I might come back with one more question, though.

Senator LUMMIS. Thank you, Mr. Chairman. One of the strategies that China has been using to increase its footprint in space is helping developing countries gain access to space. For example, in 2019, China provided satellite development and launch services to Ethiopia to launch that country's first satellite.

So, it seems to me the U.S. is generally more focused on relationships with nations that are already spacefaring. Do you see a value in the U.S. helping countries without programs access space? And how do you think we could or should do that? Mr. Bridenstine.

Mr. BRIDENSTINE. So, I will go first and then, I will turn it over to the others. So, the—when we think about the value of low Earth orbit, specifically, Mike mentioned a lot of items, as far as, you know, biotechnology and things like that. We think about pharmaceuticals, immunizations, the ability to create human tissue using your own DNA. Your own adult skin cells we can use to create your own tissue for regenerative medicine. The idea that we can use advanced materials, we can create—you know, potentially in future, we will be able to create an artificial retina for the human eyeball. So, people who have macular degeneration do not have to lose their eyesight.

So, these are things that are begin tested, right now, on the International Space Station. And when we cede that—when we cede that to China, if we do not have a replacement for the International Space Station, they are going to be able to benefit economically from all of those values of microgravity that we are not going to have access to, in the way that we currently have access to.

Now, as far as the international piece, you know, our—we have built capacity for countries in the past. We think about—well, I do not want to name any specific countries, but we have worked very hard on all of our international partners on the International Space Station. Fifteen countries operate the ISS. We have got, you know, 19 different countries that have had astronauts on the ISS. Those capacities are largely built by us.

And when we decide we are not going to have a next generation space station, which right now is the signal our country is sending by not funding the next generation human space station—I would say I saw, you know, defense—or, you know, the appropriations billfor CJS. I think it had $101 million in there for LEO commercial destinations, which I think is fantastic. But—but that is still not enough, just to be honest. We need more. We do not want to cede that territory to our biggest competitor, primarily because all of that capacity we built for these other countries is going to go to them. And they are going to start benefiting from that.

So, we have to maintain the leadership here. I think it is absolutely important. Countries that—when we think about Artemis and the Artemis Accords, we have been inviting countries that do not even have a space program, to participate in Artemis in whatever small way they can participate. Because, in my view, space is a tool of diplomacy for this country. It is something that every country wants, and we can help provide it.

So, when we think about the big picture, kind of, how the Nation thinks about agreements around the world, I think space needs to be on the table, regularly. When we—even when we do trade agreements, when we do any kind of large negotiation with another nation or many nations, space needs to be on the table because it is a great tool of diplomacy that can improve our position.

Mr. GOLD. Senator, it is a great question and even better opportunity for me to vent. The Intergovernmental agreement, the IGA, which was the agreement that established the International Space Station—it is extraordinarily difficult, if not impossible for other nations to join the IGA, which has prevented NASA from reaching out to other countries that do not have space programs or have space programs they just started.

The Artemis Accords were, in many ways, a reaction to that, that we wanted to create a vehicle that, no matter how modest your program was—even if it was just a couple of graduate students, that you could contribute to this unprecedented journey to the moon. And in the initiation of the Artemis Accords, you had countries like Luxembourg that has a relatively small but mighty space program and they are—by the way, shout out to the Embassy there at Luxembourg that has been so supportive. But also, United Arab Emirates which, again, has done amazing things in space but was not a traditional ally, certainly not on the International Space Station, yet.

So, through the Accords now, it is so important that we reach out to developing countries. I was so proud when Brazil joined the Artemis Accords, because that was our first Latin American country. And I certainly hope that we can proceed in Africa where, by the way, China has been very aggressive, in terms of international, diplomatic development. We need African countries in the Accords, supporting not just the Artemis program, but sensible norms of behavior. And I think that is our best vehicle to proceed and get these countries involved.

Senator LUMMIS. Dr. Dittmar, you were nodding.

Dr. DITTMAR. So, I would agree that it has been difficult for other countries to join the IGA, but I would also point out that the ISS has been used as a means to bring many countries, who are not spacefaring countries, into a larger community. It has, at this point, hosted—has, national projects of one type over another, for well over 105 countries, at this point, over its lifespan. And so, it has served as a tool of tremendous international diplomacy, notjust among the original 15. If I was back in—I am back, old enough back in the days that Brazil was in discussion about that, right? So, this is very nice to, sort of, see that come around again with Artemis Accords. But so, it has actually done that.

From the Axiom side, for example, as we are developing that space station, we are getting a lot of interest from internationals who see the commercial opportunity as a way to, sort of, bring themselves into the larger community of nations that are flying professional astronauts from those countries who are able to, sort of, use that. So, there is a lot of interest. I mean, we are finding that just even from a business point of view, let alone, sort of, national interest point of view.

I think if we cede this to China. I am alarmed, frankly. And I mentioned earlier that the reason I joined Axiom was I saw them as best positioned at this particular point in time, and I still believe that. But I am alarmed by what I see as the potential for a gap, which we have already talked about earlier.

And it goes back to what I said about objectives, right? I mean, in my written testimony, one of the things I said, and I think I said it orally, too, is that, you know, our objective needs to be assert and maintain U.S. leadership in low Earth orbit. Not just in low Earth orbit, but in all of space. But when we are talking about low Earth orbit that has always been done through exercise of soft power and use of soft power to bring other countries to our side. So, I think it is absolutely critical that we continue to do this.

Senator LUMMIS. Dr. Sanders, is this something you wish to address?

Dr. SANDERS. Well, I will talk about the International Space Station and just quickly reiterate a point that was made earlier about the importance of having a persistent presence in low Earth orbit. We look at it from the Panel's perspective on the risk reduction that it provides for further exploration. And if you do not have that kind of persistent presence, in order to understand the effects of low Earth orbit of microgravity and the environment on the human physiology, for example, as well as other things, than you are not going to have the ability—you are going to have greater risk when you go further away. And so, there is—there needs to be that and there needs to be a good transition period, a good transition handoff to whatever follows the ISS. And there needs to be something to follow the ISS, however we do it.

Senator LUMMIS. Thank you very much, Panel. I yield back.

Senator HICKENLOOPER. Thank you, Senator Lummis. Senator Cruz has joined us. Are you ready for your questions?

Statement of Hon. Ted Cruz, U.S. Senator from Texas

Senator CRUZ. Sure. Thank you, Mr. Chairman. Welcome to the witnesses. Good to see all of you again. Thank you for your good work.

As everyone on this panel knows, the International Space Station has been a big priority of mine for a long time. And not only is it a critical tool for projecting American leadership in space, something all the more important with the Chinese successfully putting the core module of their space station up, in April of this year, but we have also invested over $100 billion in it. And to be fiscally responsible and prudent, we need to get maximum use out of that investment.

Multiple times, the Senate has taken up legislation I have introduced to extend the ISS through 2030. The extension through 2030 was part of the NASA Authorization Act, which the Senate passed unanimously at the end of last year. It was also part of that same NASA Authorization Act that was included this year in the U.S. Innovation and Competition Act, which was passed by the Senate in May. And it is something that the current Administrator, Senator Nelson, strongly supports, as well. And so, both the current Administrator and the prior Administrator, we worked together on this extension. It is common sense. It is bipartisan. We are waiting on the House to act. And we will see if and when the House does act. But at least on the Senate side, it is 100 to nothing that this extension makes sense.

And I will mention to the Chairman, so there were battles in the prior administration because there were a few misguided voices in the Trump Administration that wanted to retire the ISS early. And Senator Nelson and I, on this subcommittee, took turns taking a two by four to the administration on that question. And it only took about 300 to nothing votes for them to realize that, perhaps their position was not going to prevail in the Senate. But we need to see the House act.

Let me ask the witnesses—look, I think we are going to get the extension done, in one vehicle or another. It is the right thing to do. It is long overdue. But given the experience of the panel, I am curious, technologically, from a safety perspective, what is keeping the station from lasting longer than 2030? From lasting through 2035 or 2040? There is a lot of expertise on this panel and so what is your collective judgment?

Mr. BRIDENSTINE. I will go ahead and start. First of all, Senator, the wounds from the two by four have recovered quite nicely and I think you are—

Senator CRUZ. To be clear, Jim, you were not the one advocating this.

Mr. BRIDENSTINE. No, I—I am aware of that but I was part of the administration. And look, here is the thing. I think what we are seeing now is that the foresight that you had at the time was right, because we are not ready for what comes after the International Space—we are not ready for it. Building a space station takes a long time, especially when you are doing it in a way that has never been done before which, in the future, will be commercial. I do not foresee Congress appropriating funds for a second International Space Station. I do not foresee that.

Now, that being said, here is the thing that I think is important. Extending it to 2030, I know the Senate has already passed that, and I support that. I would also tell you that there is no guarantee we are going to make it to 2030. Certainly, we should, if we can. We have seen already a crack on the International Space Station. We have seen a hole on the International Space Station. If you look at the outside of the space station, it has been, you know, pelted by debris. And of course, there is always modifications, upgrades that have to happen in order to keep it, you know, moving forward.

So, it is a—it is a marvel of engineering. Mike Gold used to tell me quite frequently we need to put it in for a Noble Peace Prize, which I think is right. It is a tool of diplomacy. It has been just a phenomenal capability for our country all around, not even considering how much science is coming from it right now.

So, the key is, extending it. Yes, there is a challenge. We know it cannot last forever. How far can it last? I do not think we have that answer. Right now, I think it is—we are in great shape.

Senator CRUZ. Well, and let me ask a related question, which is what do you all think is the timing for a replacement of comparable capability? And when is that transition realistically possible? And how do we ensure that the ISS is operational—that we do not cede low Earth orbit, for a period of time, to the Chinese, that we maintain U.S. leadership continuously?

Mr. BRIDENSTINE. So, the future is going to depend on how much it gets funded. Right now, it has been funded at nothing compared to what it needs. I think it was, like, $15 million, or something like that, for the transition to a new commercial, you know, space station. But what we need—and the CGS Appropriations Bill had $101 million, which met the President's budget request. That coming from the Senate—the Senate had $101 million in there to meet the President's budget request, which is fantastic. But I am telling you, sir, it is still not enough.

When we think about how long it takes to develop a space station, especially in a way that has not been done before, I do not know how long, I am not going to give you an answer on the date that that, but here is what—here is that I think the Senate should do. The Senate should absolutely declare that NASA needs to tell it when it is going to—what is the objective to have that new space station and then, the Senate needs to fund the requirements to achieve that.

I do not think the right answer is to continue—first of all, extending the space station is the right thing to do. But continuing that, in perpetuity, believing that it is going to last forever, I think is not the right approach. And I am not suggesting that is what the Senate is doing, at all.

But NASA needs to say, you know, "Here is how we are going to replace it. Here is what it is going to cost." They need to put that in the President's budget request with many years—you know, 5-year outlook, and then, come to you and say, "This is the money that we need." And right now, I do not see that happening the way it should happen.

Dr. DITTMAR. And that begins with, I think, NASA needs to completely fulfill the spirit of what was written into the 2017 NASA Transition Authorization Act, having to do with the transition plan. And a transition plan needs to have, you know, timing, milestones, clear objectives, how those objectives are going to be met. And then, once that is in place, then you can begin to have a conversation about follow-ons.

So, as you know, Axiom—because we have—you know, we have spoken to you about this, Axiom's approach, you know, was funded on a competitively awarded agreement that was negotiated with NASA. And that $101 million that Jim was talking about, when you look at how NASA is planning to allocate it, does not meet the commitment to Axiom for 2022. The work that needs to go to work on the space station side of it, in other words, for the station to do the analysis that is needed, in order to ensure that Axiom can reach orbit and dock by 2024, is not funded completely in that amount. Let alone, looking at the creation of dual path, you know, for having maybe more of an alternative to it.

So, I would agree, definitely more funding is needed. But also, NASA needs to be clear about objectives and the means that it sees at this point. Understanding that any transition plan is going to be a stake in the sand at the moment, it is going to have to be iterated upon, but it needs to be much clearer about what those objectives and milestones are.

Mr. GOLD. So, Senator, I do not know if I can enter this into the record, but I borrowed my son's ISS folder for today's testimony. I am such a fan. We talk about a lot of recycling on Earth, and I can tell you it is even more important on orbit. As you pointed out— and again, I appreciate your support and your two by four. That kept us going, you know, during some difficult times. We need to squeeze every minute and ounce of capability out of the ISS.

And in terms of when we should be retiring it—I mean, again, I am a recovering attorney but, you know, the engineers tell me there are seals that will wear out, etc. But there is still going to be good hardware and we should look at, yes, maybe there is a point of retirement, but continuing to leverage the hardware that we can continue to use as part of another system or as part of a smaller system.

And as we discussed before you arrived, you know, at Redwire Space we are the global leader of microgravity research, development, and manufacturing. The only company to actually build things on the ISS. And we are looking at a future with biotech and organs, etc., that you can build. We cannot see those technological capabilities or the diplomatic or political benefits of low Earth orbit to China. So, we need to have a two-pronged approach. One, extend the ISS to 2030, which I included in my written remarks. And then, ensure that we proceed with enough funding and capability to deploy a free-flying commercial space station so that it is operational while the ISS is still there, and we do not create a gap in low Earth orbit that would be disastrous for us and the Nation. And as you point out, you have fans at NASA in doing that. You know, Senator Nelson—Administrator Nelson has been a great supporter and we have talked a lot about the need for an authorization bill. You have former staff that was working on that authorization bill that is now over at NASA. So, I think you have got great allies at the Agency.

Senator CRUZ. Thank you.

Senator HICKENLOOPER. Great. Thank you, Senator. Appreciate you making the effort to get over here. I think we are at that point where I am forced by tradition and circumstance to let you go.

The hearing record will remain open until November 18, 2021. Any Senators who would like to submit questions for the record, for the witnesses, should do so by November 4, 2021. To all of you witnesses, in addition to my gratitude, we ask that your responses be returned to the Committee by November 18, 2021.

Again, I would like to thank each of you for your testimony today, but really, your service in terms of maintaining or helping to create America's role as a leader in space, but also to maintaining.

That concludes our hearing for today. We are adjourned. [Whereupon, at 12:03 p.m., the hearing was adjourned.]

Appendix

Response to Written Questions Submitted by Hon. Kyrsten Sinema to Dr. Mary Lynne Dittmar

NASA Reauthorization. The U.S. Innovation and Competition Act (USICA) includes legislative language to reauthorize the National Aeronautics and Space Administration (NASA) that I developed on a bipartisan basis last year with Senators Cantwell, Wicker, and Cruz. NASA was last authorized in 2017.

Question 1. Why is it important for Congress to reauthorize NASA after failing to do so for many years?

Answer. It is important and necessary for Congress to pass authorization acts for several reasons. Authorization laws are intended to create, continue, and modify programs. These bills provide policy guidance and ensure that the goals of an age cy are met. For NASA, it is critical that constancy of purpose is prioritized through authorization acts, so that the agency's multi-year investments and programs can maintain stability with regard to schedule and cost, and thus, support the agency as it strives to meet its goals. Authorization acts also send a signal to the international community of the intent of Congress to continue to assert U.S. leadership in space—a signal that is important not only to our allies, but also to those who do not wish us well. Finally, authorization acts provide a critical tool for Congressional oversight.

Question 2. Do you support the reauthorization language in the USICA bill? Answer. There are many strong components of the Senate-passed bill. As I mentioned in my testimony before the committee, I applaud Congress for including language (once again) that it is the priority of the United States to maintain a continuous human presence in LEO (section 2620 (1–3, 3(b)). In light of the findings and recommendations in the DoD Defense Innovation Unit's 'State of the Space Industrial Base" released on November 18, Congress may consider further strengthening the language to read "continuous American presence in space, beginning with low Earth orbit." The report can be found by accessing the linked pdf below.

https://assets.ctfassets.net/3nanhbfkr0pc/43TeQTAmdYrym5DTDrhjd3/a37eb4fac2bf9add1ab9f71299392043/Space_Industrial_Base_Workshop_2021_Summary_Report_-_Final_15_Nov_2021c.pdf.

In addition, I support the reauthorization language that extends the ISS to "at least 2030" (Section 2521, and conforming amendments. Beyond 2030, the costs of maintenance, operations, and particularly the engineering lifecycle data should be evaluated carefully, as well as the ongoing role of the ISS.

Certainty is required by commercial providers who are developing follow on stations to close their business cases which in turn is imperative to meet the higher order requirement to ensure continuing U.S. interests and human presence in low Earth orbit after the ISS end- of-life.

However, regarding space suits (Section 2616, (1))), I would suggest that driving NASA to partner "to the maximum extent practicable with industry-proven spacesuit design, development and manufacturing suppliers" may limit innovation. Partnering with such entities is a good idea but so is partnering with entities that "credibly demonstrate the capability to meet requirements as determined by NASA." To that end, I would also suggest that rather than providing NASA the option to partner with "1 or more private entities," this should be changed to "2 or more," based on precisely the same logic that the re-authorization employs in its position that the Human Landing System (HLS) should involve 2 or more providers.

Regarding Section 2627, "Payments Received for Commercial Space-Enabled Production on the ISS," I commend the committee for placing this into context by directing the NASA Administrator to take into account ''the relative maturity and profitability'' of the applicable product or service. However, ''profitability'' of a prod- uct or service may or may not be a simple metric, depending upon how the development of that product or service has been funded by a company. Most importantly, however, the low Earth orbit ''economy'' does not yet exist; it is a nascent entity that at its best creates goods and services that are executed in part in space but are based wholly on Earth. Congress should be doing everything in its power to enable companies to demonstrate profitability in production efforts—a goal that is still a long way off—to encourage investors to come to the table to support them and most importantly to support the commercial space stations that Congress intends should eventually take the place of the ISS. Those companies will be the customers of such stations and the entities trying to raise capital now on the private market must be able to assure their investors that business cases can close by 'going concerns' that have time to fully demonstrate their viability on the ISS without government levying a "lien" of sorts that will then be deposited into a government-managed space exploration fund.

Beyond the specific language of NASA reauthorization language, I also strongly support the language regarding Space Situational Awareness and encourage Congress to fund this effort and give it full attention and encouragement. Last week's Russian ASAT test illustrates (again) the need for U.S. leadership and orchestration of the complex issues related to SSA. It is critical that congress work with the Administration to confirm a Director for

OSC who will provide clear leadership to OSC and provide a focal point for interagency and across-industry collaboration. Further, OSC should be elevated into a separate bureau—the Bureau of Space Commerce— instead of being left within the National Oceanic and Atmospheric Administration (NOAA). OSC should be given authority over clear goals, namely: promoting commercial space and championing it within the government, aiding in the development of standards and norms of behavior in space, identifying and addressing regulatory gaps. Moving OSC out from NOAA will give it the independence to establish a new, clean cut, directive organizational culture.

International Space Competition. As discussed in the hearing, the United States faces growing international competition in space. The Chinese government is building its own space station, recently launched a hypersonic missile, and is spending billions of dollars on establishing and promoting new space capabilities that crossover from military to civil applications and back again ('military civil fusion, or MCF'). In addition, companies in the United States and several abroad have applied for tens, or in some cases, hundreds of thousands of satellites in low-Earth orbit.

Question. What actions should Congress take to ensure that NASA and the United States remain global leaders in space?

Answer. It is undeniable that China is moving fast in LEO, and forging partnerships with members of the international community for activities onboard their station and in cislunar space. In my testimony, I noted that China is playing the long game, and it is critical that in assessing China's intentions in space, Congress understand how the People's Republic of China (PRC) is advancing rapidly across a range of space technologies and capabilities. The U.S.-China Economic and Security Review Commission (USCC) produced a report to Congress in 2019 that outlined the need for a comprehensive, wholistic perspective regarding China's competitive interests in space. In my testimony, I included the USCC's findings on China's strategy to catch up and overtake the U.S. in preeminence in space.

In the National Defense Authorization Act of 2020, Congress responded to the USCC report by codifying recommendations for action. Namely, the defense bill called for the President, in consultation with the National Space Council, to develop ''. . . a strategy to ensure the United States can effectively compete with other national space programs, maintain dominance in the emerging commercial space economy, and has market, regulatory, and other means available to address unfair competition by the PRC . . .'' and also "a plan to strengthen and streamline cooperation with international allies and partners in space."[26]

NASA, and by extension U.S. space companies serve as a vehicle of U.S. "soft power," shaping favorable behavior and outcomes among foreign entities, customers, and over time, competitors who do not operate according to American laws and values. China's ability to plan and make investments 5–10 years out under a militaryrun program and an increasingly totalitarian government is intended to enable Chinese hegemony in space with a fairly high degree of certainty. This, unfortunately, is an area where the U.S. must look itself in the mirror and understand that it must dedicate itself to constancy of purpose and multi-year strategic planning and investment. We have maintained our leadership role in space for many years thanks to the great public servants of NASA and ingenuity of the American people, but China is making great strides in their goal to meet pace with America. If the U.S. Congress hopes to maintain its leadership role in low-Earth orbit, for example, a forward thinking and thoughtful plan for a post-ISS world must come sooner than later. Our international partners, U.S. commercial partners, and the scientific community must be engaged as Congress strategically positions the U.S. to continue thriving and exploring space.

To go further, Congress must recognize that space as a domain extends far beyond NASA's remit. Congressional committee structure makes it difficult to meaningfully define strategy for the U.S. future in space, which is in many ways the same thing as the U.S. future and leadership on Earth. Space is simultaneously a contested and crowded corridor for transportation of goods, services, and assurance of security, an economic engine promoting job growth, an operational domain for civil, military, and commercial entities, a key component of U.S. infrastructure, a collaborative and competitive zone for foreign powers, a security-critical 'location', a driver of technology advancement, plays a key role in monitoring and possibly mitigating climate change, and much more. As pointed out in the aforementioned DIU publication, the state of the U.S. space industrial base is "tactically strong but strategically fragile." In my opinion the U.S. is not prepared for the increasing competition in space. Becoming prepared will require a "whole of government" approach that Congress has not yet evidenced, beginning with addressing institutional barriers in committee scope, collaboration, and oversight. Full utilization of a bipartisan bicameral "space caucus" would be

[26] U.S. Congress (2019). National Defense Authorization Act of 2020, Pub. Law 116–92 (20 December). https://www.congress.gov/116/plaws/publ92/PLAW–116publ92.pdf.

a start, as would work on joint hearings, for example on the establishment of an "economic security corridor" extending from the Earth to the Moon (and beyond). Another approach would be to extend support for NASA's Artemis Accords by authorizing both Artemis and the Accords and advancing their international goals, as the Committee just did in this most recent hearing.

Role of University Partnerships. Much of the hearing discussion focused on the International Space Station and near-earth orbit satellites. Arizona universities support these missions, and additionally, the universities are leaders in other NASA projects. For example, Arizona State University's OSIRIS–REx Thermal Emission Spectrometer has collected samples from an asteroid and will return to Earth in 2023. Meanwhile, the University of Arizona's Near-Earth Objection (NEO) Surveyor Mission will help NASA complete its Congressional mandate to discover 90 percent of all near-Earth objects larger than 140 meters in length.

Question 1. How can we ensure that crucial science and planetary defense missions, such as OSIRIS–REx and the NEO Surveyor, receive the support from NASA to be successful?

Answer. NASA's Human Exploration programs and its Space Science missions are two sides of the same coin; the human aspiration and intention to inquire of the universe "Where did we come from? Where are we going?" Our understanding of the universe is advanced step-by-step through robust support and decades-long development and funding of space science programs such as OSIRIS-Rex and the NEO Surveyor. However, these particular missions play a dual role; that of generating knowledge needed to develop the systems we need to protect our planet from NEO incursions. Critical programs such as OSIRIS-Rex and the NEO Surveyor need champions in Congress who will continue to support these and other missions through strong authorization and appropriations language. This can be difficult in the face of many other pressing—and apparently more immediate—national priorities; however, the question of planetary defense is only a question until it's not.

Question 2. In your opinion, how can NASA best utilize university partnerships to perform research, advance science, and help NASA reach its goals?

Answer. University partnerships are important for engaging the next generation of science and engineering leaders across the United States, and for expanding both the nature and quantity of collaborative partnerships related to NASA's mission and extending that mission. For example, NASA's L'SPACE Academy, a collaborative partnership between NASA and Arizona State University (ASU), provides the opportunity for undergraduate students to participate in free, online, and interactive programs that encourages students to learn about NASA missions and connects them with industry professionals. https://www.lspace.asu.edu/ In addition, ASU is the lead university for the Psyche Mission—an extraordinary mission to a 'metal asteroid'. The Psyche team in turn has set up an online course about the mission which encourages participation and learning not only about Psyche but in related STEM fields. ASU was also the 2020 winner of NASA's Big Idea Challenge—https:// bigidea.nianet.org—which provides opportunity for partnerships with universities and other entities. Finally, competitions run by NASA that provide opportunities for universities to collaborate with the agency and with other companies has been a part of NASA's rich legacy of stimulating both basic and applied research since the earliest days of the space program. Finally, ASU is also a partner in the MILO Space Institute—an innovative partnership between ASU and Lockheed Martin to establish momentum toward public-private partnerships in science. Not to be overlooked, Arizona University is a key contributor to pathfinding NASA missions such as the Phoenix Mars Lander, the James Webb Space Telescope, which will enable humanity to peer back through time to the earliest moments after the beginning of the universe, and OSIRIS-Rex.

Arizona's Apollo History. During the Apollo mission in the 1960s and 1970s, NASA astronauts trained at Sunset Crater National Monument near Flagstaff, Arizona, where the volcanic landscape approximated the lunar surface. Astronauts also trained at Meteor Crater in the high desert of northern Arizona.

Question. Could these locations in Arizona be beneficial to Artemis astronauts as they train for their lunar missions?

Answer. Training for spaceflight missions is informed by all that has been learned before, from previous missions and training programs. If the locations are still deemed to be close approximations of the lunar surface—in particular the environment around the south pole of the Moon, where the Artemis missions are intended to be carried out—then they would certainly be reasonable candidates for training.

Response to Written Questions Submitted by Hon. Raphael Warnock to Dr. Mary Lynne Dittmar

Regulatory Framework for Commercial Spaceflight. In recent years, we have seen the tremendous growth in the commercial space sector. As multiple witnesses have noted, the responsibility of regulating this industry is spread across a variety of Federal agencies, with differing levels of technical expertise and industry outreach.

Question 1. How can Congress improve interagency cooperation to better serve America's commercial space sector?

Answer. The Office of Space Commerce (OSC) is the designated lead agency for representing commercial space sector interests in interagency discussions, yet the office's director position remains vacant. It is critical that Congress work with the Administration to confirm a Director for OSC who will provide clear leadership to the function of OSC.

Further, OSC should be elevated into a separate bureau—the Bureau of Space Commerce—instead of being left within the National Oceanic and Atmospheric Administration (NOAA). OSC should be given authority over clear goals, namely: promoting commercial space and championing it within the government, aiding in the development of standards and norms of behavior in space, identifying and addressing regulatory gaps. Moving OSC out from NOAA will give it the independence to establish a new, clear cut, directive organizational culture.

Once OSC is elevated and established in its roles and responsibilities, interagency cooperation will be made easier with the benefit of jurisdictional clarity.

Question 2. Do you believe that Congress Federal regulatory oversight of commercial space endeavors should be more concentrated within one agency than exists under the current regulatory landscape? If so, what are the considerations that Congress should keep in mind as it designates or establishes such an office?

Answer. There are benefits to different agencies regulating certain aspects of commercial space. FAA's Office of Commercial Space Transportation (AST), guided by Chapter 509 of Title 51, ensures the safety of the uninvolved public and their property, protects national policy goals, and enables industry growth by providing guidance, launch licenses, experimental permits, and promotional support including R&D. FAA AST has developed considerable expertise in commercial space transportation and has done a good job meeting the challenges that come with a rapidly growing industry. FAA AST requires continued oversight from Congress to balance the needs of industry to develop without premature regulatory burdens with the need to ensure, as far as possible, public safety.

The Office of Space Commerce (OSC) has been tasked with shaping civil space situational awareness, modernizing oversight and regulation activities, and promoting the U.S. civil space industry internationally. This office's role is more important than ever—with a growing debris field around the earth, posing a serious threat to unfettered access to space, and our national assets that exist in the space domain— many of which contribute in a critical way to U.S. national security and our day- to-day activities. In addition, in the recent past OSC has played an important "convening" role for the commercial space industry, encouraging a free exchange of ideas, and building cross-agency bridges through sponsorship of events with the Department of State and the U.S. Trade Organization. These events and exchanges are an important part of the government function and a way that the government can encourage industry growth.

Question 3. How should Federal agencies best engage with industry to ensure that Federal regulation of commercial space entities meets the need for public safety and oversight without unnecessarily curbing innovation?

Answer. The U.S. Commercial Space Launch Act of 1984 as amended (51 USC 50901) is the primary law guiding Federal oversight, regulation, and promotion of the U.S. commercial space transportation industry. The law mandates that the Secretary of Transportation ensure that all launch and reentry activities shall protect public safety of property and support U.S. national security and policy interests, and that the Secretary shall issue licenses to U.S. applicants who show that they do and will continue to meet safety requirements. The CSLA has been amended several times since 1984. Critical to the regulatory regime's enabling nature is its exclusive focus on protecting the uninvolved public and their property. Space transportation continues to be seen as a dangerous and risky activity, i.e., not a common carrier mode of transportation with an expectation of safety and success for those who opt to participate. Thus, in the case of human spaceflight, the law expressly limits regulation for their safety and calls for informed consent of the flight participants. The intention of this "learning period" is to allow the industry time to grow and develop best practices ahead of the eventual arrival of regulatory action by the government. The commercial space industry in the United States is dynamic and evolving. A great example of this growth can be seen by the increase in launches licensed by the FAA per year. In 2011, the FAA licensed only one launch, in 2016, they licensed 11, and in 2021 so far, 47 licensed launches have occurred.[27] While the industry has certainly experienced immense growth over the past decade, it would be unrealistic to

[27] "FAA Licensed Launches." Licensed Launches. Federal Aviation Administration, April 19, 2021. https://www.faa.gov/data_research/commercial_space_data/launches/?type=license.

determine it deserves the same type of regulatory landscape as that of commercial aviation, which transportations thousands of people globally each day. Given the technical and economic challenges in rocket science, this bounding of regulatory risk—while still protecting American citizens and interests from harm—gives entrepreneurs and investors' confidence that they will get permission to fly.

Continued encouragement by Congress on the development of safety standards and best practices between industry and the government will lead to an informed and mutually beneficial relationship for all parties involved, while not prematurely hamstringing the growth and progress of the commercial space sector in a time where America sees foreign powers rapidly developing this critical technology of their own.

FAA/AST faces a growing workload, as launch cadences continue trending upward. As mentioned, AST requires the resources to meet this exciting challenge. Furthermore, Congress should invest directly in the consensus standards effort identified by FAA's Commercial Space Transportation Advisory Committee (COMSTAC) as the agreed-on consensus choice for human spaceflight standards- writing.

Response to Written Questions Submitted by Hon. Kyrsten Sinema to Mike Gold

NASA Reauthorization. The U.S. Innovation and Competition Act (USICA) includes legislative language to reauthorize the National Aeronautics and Space Administration (NASA) that I developed on a bipartisan basis last year with Senators Cantwell, Wicker, and Cruz. NASA was last authorized in 2017.

Question 1. Why is it important for Congress to reauthorize NASA after failing to do so for many years?

Answer. Since Apollo, no NASA beyond low Earth orbit ("LEO") human spaceflight program has come to fruition, leaving our international partners justifiably skeptical of the Agency's ability to sustain such efforts and achieve success. This stands in stark contrast to China which has maintained both consistency and continuity for their own beyond LEO human spaceflight efforts and their space program generally. The best way to reassure our partners in Europe, Japan, Canada, and throughout the world is for Congress to pass a bipartisan reauthorization bill that demonstrates uniform support for the Artemis program and key aspects such as the Lunar Gateway which international partners are contributing billions of dollars to.

Question 2. Do you support the reauthorization language in the USICA bill?

Answer. Yes, the language in the USICA provides critical support for Artemis as well as for other important issues such as the extension of the International Space Station (ISS) to 2030 and ensuring the development of a commercial space station to ensure that the U.S. maintains a constant crewed presence in LEO.

International Space Competition. As discussed in the hearing, the United States faces additional international competition in space. The Chinese government plans to develop its own space station and recently launched a hypersonic missile. Other nations have applied for tens, or in some cases, hundreds of thousands of satellites in low-Earth orbit.

Question. What actions should Congress take to ensure that NASA and the United States remain global leaders in space?

Answer. Again, Congress should pass a bipartisan authorization bill that explicitly supports the Artemis program. Moreover, Congress should extend the ISS until at least 2030 and direct NASA to support a future commercial space station by serving as a robust customer for such a platform including the purchase of seats not just for U.S. astronauts but for international partners as well allowing the Agency to maintain its traditional barter system and leadership role in space exploration. Congress should also take action to ensure that the U.S. does not fall behind China and other nations in orbital servicing, assembly, and manufacturing. Specifically, Congress should explicitly direct NASA to support critical programs such as OSAM–3 and the deployment of an Earth Science Platform which will bolster our knowledge of climate change while developing critical new capabilities to ensure that the U.S. doesn't fall behind China in this vital field of space-based robotics/next-generation persistent platforms.

Role of University Partnerships. Much of the hearing discussion focused on the International Space Station and near-earth orbit satellites. Arizona universities support these missions, and additionally, the universities are leaders in other NASA projects. For example, Arizona State University's OSIRIS–REx Thermal Emission Spectrometer has collected samples from an asteroid and will return to Earth in 2023. Meanwhile, the University of Arizona's Near-Earth Objection (NEO) Surveyor Mission will help NASA complete its Congressional mandate to discover 90 percent of all near-Earth objects larger than 140 meters in length.

Question 1. How can we ensure that crucial science and planetary defense missions, such as OSIRIS–REx and the NEO Surveyor, receive the support from NASA to be successful?

Answer. Congress should include explicit direction in authorization and appropriations bills directing NASA to leverage Archinaut-based technologies to enhance the capabilities of science and planetary defense missions while lowering their costs.

Universities also have an important role to play in support of new commercial space stations. Specifically, Redwire Space is proud to be a partner with Arizona State University in support of the Orbital Reef, a next-generation commercial space station. Other Orbital Reef team members include Blue Origin, Sierra Space, and Boeing.

Question 2. In your opinion, how can NASA best utilize university partnerships to perform research, advance science, and help NASA reach its goals?

Answer. NASA should leverage universities to support key programs and missions, taking advantage of educational institutions' robust capabilities and to support vital workforce development to ensure that the U.S. does not fall behind China or other nations in space development and operations.

Arizona's Apollo History. During the Apollo mission in the 1960s and 1970s, NASA astronauts trained at Sunset Crater National Monument near Flagstaff, Arizona, where the volcanic landscape approximated the lunar surface. Astronauts also trained at Meteor Crater in the high desert of northern Arizona.

Question. Could these locations in Arizona be beneficial to Artemis astronauts as they train for their lunar missions?

Answer. Absolutely. Continued ground training in Arizona and in other states will be a critical part of ensuring that Artemis astronauts are prepared to tackle the numerous challenges they will face exploring the Moon and Mars.

Response to Written Questions Submitted by Hon. Raphael Warnock to Mike Gold

Regulatory Framework for Commercial Spaceflight. In recent years, we have seen the tremendous growth in the commercial space sector. As multiple witnesses have noted, the responsibility of regulating this industry are spread across a variety of Federal agencies, with differing levels of technical expertise and industry outreach.

Question 1. How can Congress improve interagency cooperation to better serve America's commercial space sector?

Answer. Thank you for asking such an important question. Establishing a proper regulatory framework which balances innovation and commercial development with preserving the space environment and meeting our international treaty obligations is vital. The most glaring issue that needs to be addressed is to establish an explicit system for providing ''continuing supervision'' as required by the Outer Space Treaty of 1967 for commercial activities including private sector space stations; orbital servicing, assembly, and manufacturing; commercial lunar rovers; and mining activities on the Moon or asteroids.

Congress should designate an entity within the Federal Government to exercise this continuing supervision activity by creating a 'Mission Authorization' process that would require companies to describe their proposed activities prior to launch and subsequently update the government of any substantive changes. This would provide a benign means of meeting the 'continuing supervision' requirement that wouldn't unduly burden entrepreneurs while still generating the information necessary that, per the requirements of the Artemis Accords, would be publicly shared with the United Nations to prevent confusion and conflict in space.

Question 2. Do you believe that Congress Federal regulatory oversight of commercial space endeavors should be more concentrated within one agency than exists under the current regulatory landscape? If so, what are the considerations that Congress should keep in mind as it designates or establishes such an office?

Answer. Further concentration of regulatory oversight for all commercial space endeavors into a single agency would be ideal. However, for such a paradigm to succeed, this single agency not only needs the proper authority, but must also be properly resourced in terms of funds and experienced personnel to avoid causing undue burdens that would harm American competitiveness, job growth, and scientific activities such as combating climate change.

Question 3. How should Federal agencies best engage with industry to ensure that Federal regulation of commercial space entities meets the need for public safety and oversight without unnecessarily curbing innovation?

Answer. All Federal agencies that engage in regulating commercial space activities should create or leverage existing Federal Advisory Committees to ensure that industry and government are acting in concert with each other to maximize both innovation and safety. Government and industry are stronger when they're working together. For example, during the hearing a question was asked about the NASA Advisory Council's Regulatory and Policy Committee (NAC RPC). This Committee is comprised exclusively of industry representatives and is vital to ensure that NASA and the private sector are communicating with each other on important regulatory and policy issues. In stark contrast, the NAC itself has only one industry representative which is completely at odds with the large and important role private sector entities play in executing NASA's mission. NASA should proceed with appointing more industry representatives to the NAC and its various committees, re-double its use of the NAC RPC, and other agencies/departments, such as the Departments of State and Commerce, should develop their own Federal Advisory Committees focused on commercial space activities.

Response to Written Questions Submitted by Hon. Kyrsten Sinema to Dr. Patricia Sanders

NASA Reauthorization. The U.S. Innovation and Competition Act (USICA) includes legislative language to reauthorize the National Aeronautics and Space Administration (NASA) that I developed on a bipartisan basis last year with Senators Cantwell, Wicker, and Cruz. NASA was last authorized in 2017.

Question 1. Why is it important for Congress to reauthorize NASA after failing to do so for many years?

Answer. The space environment is changing rapidly and is not the same as it was in 2017. Reauthorization of NASA at this time would be advantageous to set expectations for the Agency in the current and emerging environment.

Question 2. Do you support the reauthorization language in the USICA bill?

Answer. The USICA bill is very complex and wide-ranging. I would only comment on Part II which addresses space and only in general terms relevant to the charter of the ASAP. The Panel is pleased to see the emphasis placed on Space traffic Management and Space Situation Awareness which is consistent with the recommendation we made in 2020. We also support, as important safety and risk management approaches, the emphasis on continuing presence in Low Earth Orbit, the support for competing designs for the Human Landing Systems, the need for work on advanced space suits, and a step-wise approach to space exploration.

One issue not addressed in the USICA bill is the much needed revision of the language in the NASA Authorization Act of 2005 concerning Human Space Flight Independent Investigations. The ASAP noted, as far back as 2015, that the language, while perhaps appropriate and logical for the ISS and Space Shuttle, was inadequate and obsolete for the emerging environment that included reusable and commercially provided space vehicles. In 2015, the ASAP made the following recommendation:

The 2005 Authorization language should be reviewed with today's systems in mind. . .including details on the level of vehicle damage requiring investigation, the temporal issues of when mission phases begin and end, and NASA's oversight role in mishap investigations conducted by its providers, as well as when the need for oversight is required. The mishap response procedures should be thought through, documented, and in place well before any actual flights.

The 2005 Authorization language is even more inadequate when considering the emerging space travel involving non-NASA participants. Yet this recommendation remains open and increasingly relevant.

International Space Competition. As discussed in the hearing, the United States faces additional international competition in space. The Chinese government plans to develop its own space station and recently launched a hypersonic missile. Other nations have applied for tens, or in some cases, hundreds of thousands of satellites in low-Earth orbit.

Question. What actions should Congress take to ensure that NASA and the United States remain global leaders in space?

Answer. Congress can support NASA and United States global leadership by (1) providing adequate and stable resources to support a consistent program of work and a responsible and realistic schedule for that work, and (2) providing an appropriate regulatory framework to support the entire national space environment, including government and commercial entities.

Role of University Partnerships. Much of the hearing discussion focused on the International Space Station and near-earth orbit satellites. Arizona universities support these missions, and additionally, the universities are leaders in other NASA projects. For example, Arizona State University's OSIRIS–REx Thermal Emission Spectrometer has collected samples from an asteroid and will return to Earth in 2023. Meanwhile, the University of Arizona's Near-Earth Objection (NEO) Surveyor Mission will help NASA complete its Congressional mandate to discover 90 percent of all near-Earth objects larger than 140 meters in length.

Question 1. How can we ensure that crucial science and planetary defense missions, such as OSIRIS–REx and the NEO Surveyor, receive the support from NASA to be successful?

Answer. The ASAP has not focused on science and planetary defense missions, but recognizes their importance. Support and resources from the Congress will be critical to NASA support.

Question 2. In your opinion, how can NASA best utilize university partnerships to perform research, advance science, and help NASA reach its goals?

Answer. NASA can best achieve its goals by taking advantage of the full range of capabilities available to it, including those from universities, commercial industry and international partners.

Arizona's Apollo History. During the Apollo mission in the 1960s and 1970s, NASA astronauts trained at Sunset Crater National Monument near Flagstaff, Arizona, where the volcanic landscape approximated the lunar surface. Astronauts also trained at Meteor Crater in the high desert of northern Arizona.

Question. Could these locations in Arizona be beneficial to Artemis astronauts as they train for their lunar missions?

Answer. I am not sufficiently familiar with the specific locations mentioned to comment.

Response to Written Questions Submitted by Hon. Raphael Warnock to Dr. Patricia Sanders

Regulatory Framework for Commercial Spaceflight. In recent years, we have seen the tremendous growth in the commercial space sector. As multiple witnesses have noted, the responsibility of regulating this industry are spread across a variety of Federal agencies, with differing levels of technical expertise and industry outreach.

Question 1. How can Congress improve interagency cooperation to better serve America's commercial space sector?

Answer. A number of Agencies today engage with or otherwise have an influence on the commercial space sector to include: NASA, DoD, DOT, FAA, DOC, FCC and the State Department. Interagency cooperation on space issues has generally been positive. Therefore, it may not be necessary for Congress to be concerned about improving interagency cooperation.

Interagency cooperation has not been an area of focus for the ASAP, but some actions that we do believe Congress could take to better serve the United States commercial space sector—and NASA—would be to

- Avoid programmatic gaps and delays by providing adequate and timely resources for government-funded space programs. A long-term perspective is needed for since these systems can take years to develop. Following a normal appropriations process rather than adopting Continuing Resolutions would be a key step.
- Ensure that roles and responsibilities for space-related departments and agencies are clear and do not have any inappropriate overlaps or holes.
- Update the outdated, existing procedure for investigating human space flight mishaps (as previously recommended by the ASAP).
- Identify a lead Federal agency for Space Traffic Management and provide the necessary authority, immunity from lawsuits, and resources to do the job (as also previously recommended by the ASAP).

Question 2. Do you believe that Congress Federal regulatory oversight of commercial space endeavors should be more concentrated within one agency than exists under the current regulatory landscape? If so, what are the considerations that Congress should keep in mind as it designates or establishes such an office?

Answer. The ASAP has not specifically considered whether oversight of commercial space endeavors should be concentrated under one agency. It is possible that consolidation and streamlining could be accomplished by transferring functions from other organizations to one, perhaps significantly simplifying the approval process for space operators. Alternatively, or as an interim step, it may be possible to streamline the regulatory process without making major changes to responsibilities, staffing and budgets, by designating a new or existing agency to serve as the regulatory "front door" for commercial space operators. This office might then work to consolidate and streamline the needed regulatory approvals and a consistent government response.

Question 3. How should Federal agencies best engage with industry to ensure that Federal regulation of commercial space entities meets the need for public safety and oversight without unnecessarily curbing innovation?

Answer. One possible approach would be for Congress or the White House to establish top-level guidance to do just that—to ensure public safety and to encourage, facilitate, and promote commercial space transportation. As another step toward ensuring that a regulatory agency does not curb innovation, it is important to provide adequate resources for staffing with the appropriate expertise and experience to keep pace with advances in technology.

Response to Written Questions Submitted by Hon. Shelley Moore Capito to Dr. Patricia Sanders

Question 1. As you are probably familiar with, Fairmont, WV is home to the Katherine Johnson Independent Verification and Validation (IV&V) facility. The Katherine Johnson IV&V Center provides NASA the highest achievable levels of safety and cost-effectiveness for mission critical software and has the capabilities for so much more. I have spoken a lot on this issue with Former Administrator Bridenstine and with current Administrator Nelson. I believe West Virginia has the facilities, like the Katherine Johnson IV&V Facility, and the people who can be the backbone for the Artemis program. As chair of the NASA's Aerospace Safety and Advisory Panel (ASAP), could you speak to the value of the IV&V to NASA's mission.

Answer. Independent Verification and Validation is important to all software development. This is especially true for all of NASA's safety critical software.

Question 2. I am proud of the capabilities housed at the Katherine Johnson IV&V center, however I hear too often that they feel underutilized. Is that the case?

Answer. The Panel is aware that NASA employs the capabilities of that facility for all software developed by NASA and frequently for software developed by NASA contractors. We are not cognizant if that fully employs the full capacity of the Center.

Chapter 3

NASA Exploration Plans: Where We've Been and Where We're Going*

Committee on Commerce, Science, and Transportation

Washington, DC.

The Subcommittee met, pursuant to notice, at 3:04 p.m. in room SD–562, Dirksen Senate Office Building, Hon. Ted Cruz, Chairman of the Subcommittee, presiding.

Present: Senators Cruz [presiding], Sinema, Capito, and Gardner.

Opening Statement of Hon. Ted Cruz, U.S. Senator from Texas

Senator CRUZ. This hearing is called to order. Welcome.

Fifty years ago, exactly one week from today, at approximately 9:30 a.m., three astronauts, sitting atop a rocket the size of a Navy destroyer packing seven and a half million pounds of thrust, took off from Kennedy Space Center in Florida. Roughly a million people had gathered on the ground to watch this historic event, including half of the U.S. Congress.

These three astronauts, as one of the newspapers put it at the time, carried with them the hopes of the world. The year was 1969, the year before I was

---

* This is an edited, reformatted and augmented version of Hearing Before the Subcommittee on Aviation and Space of the Committee on Commerce, Science, and Transportation, United States Senate, Publication No. S. HRG. 116–600, dated July 9, 2019.

In: The National Aeronautics and Space Administration's Role …
Editor: Brent M. Jenkins
ISBN: 979-8-89530-539-3

born. The astronauts were Neil Armstrong, Buzz Aldrin, and Michael Collins, and the mission was Apollo 11. Armstrong and Aldrin would go on to make history a little more than a hundred hours later when, with more than a third of the Earth watching or listening live, they became the first humans to ever set foot on the Moon.

The Apollo 11 mission would go on to make history again, a little less than 100 hours after that, as the first mission not only to put men on the Moon but to bring them home safely, as well.

Although President Kennedy hadn't lived to see it, the bold goal that he had set 8 years earlier had been met. To steal a line from the flight director of that mission, "We had shown that what America will dare, America will do."

Today, we rightfully celebrate the momentous occasion that is the upcoming 50th Anniversary of Apollo 11. As President Nixon said in a phone call to Neil Armstrong and Buzz Aldrin while they were still on the Moon, because of what they had done, "The heavens had become a part of man's world."

Indeed, not only did we succeed in putting men on the Moon and returning them safely to Earth, but we've gone on to put robotic rovers on distant planets, celestial observatories in orbit that can literally peer into the beginnings of the universe, and we've established a presence in low-Earth orbit that is still there today.

However, while it's tempting to focus only on the historic achievements that were Apollo 11, as some of our witnesses today will rightly highlight, the Moon landing and the entire Apollo Program, for that matter, didn't happen in a vacuum.

It was the result of visionary leadership, national unity, and old-fashion American tenacity. The success of Apollo 11 and our national space program was also due in large part to the tireless contribution of countless women who were working behind the scenes and whose stories have only recently become household names.

One of our witnesses today, Dr. Christine Darden, was one of the famed human computers at NASA. Without her work and the work of other computers, many of them African American women, we never could have sent astronauts into space, let alone brought them home safely.

Unfortunately, at the time Dr. Darden and the other human computers' contributions were hidden and they remained hidden for far too long, relegated to the background.

After the movie Hidden Figures came out, a wonderful, wonderful movie that I recommend to everyone, I introduced legislation to rename the street in front of the NASA Headquarters as Hidden Figures Way.

The D.C. City Council in turn took up the idea. Just a few weeks ago, I was proud to join Dr. Darden and the family of those other legendary human computers at the dedication of the new street sign in front of the NASA Headquarters, so that a decade, or a generation, or a century from now, when a little girl or little boy goes to visit NASA, she or he will say who were they? Tell me their story.

As we look at the space landscape today, we see it's far different from the landscape of 1969. America and the Soviet Union are no longer the only players in space. Government space programs are no longer the only game in town, and our technological capabilities, both in terms of our ability to plan mission and how long these missions are, have changed dramatically.

What will the next 50 years of space exploration look like, and what should we seek to accomplish? We need a bold vision, a vision that sees the commercial space industry thriving. I've long said that the first trillionaire, I believe, will be made in space.

In 50 years, we will have gone back to the Moon. Indeed, the United States will return to the Moon as part of the Artemis Program. Artemis is the twin sister of Apollo, and this time when we return to the Moon, NASA has committed that we will land the first woman on the Moon, an American astronaut, and on behalf of my two young daughters, let me say thank you and it's about time.

From there, we'll move toward having a more permanent and sustainable presence on the Moon and then ultimately to Mars. Just a couple of years ago, I was proud to author the bipartisan NASA Authorization Act signed into law in which every Member of Congress in the House and Senate and both parties united to say that the objective of space exploration for NASA is to go to the red planet and land on Mars, and that the first boot to set foot on the surface of Mars will be that of an American astronaut.

The next 50 years have the potential to be even more consequential than the last. That's why I'm glad to be engaged with Ranking Member Sinema, with Chairman Wicker, with Ranking Member Cantwell, on yet another NASA Authorization Act to help continue to lay out a bold visionary agenda for NASA and manned space exploration so that America continues to lead the world in exploring space and exploring the great frontiers above us.

With that, I recognize Senator Sinema.

Statement of Hon. Kyrsten Sinema, U.S. Senator from Arizona

Senator SINEMA. Well, thank you, Chairman Cruz, for holding this hearing today.

As we approach the 50th Anniversary of Apollo 11, it is timely to look back at our country's accomplishments in space. It's also important for us to look ahead at new strategies and technologies that will maintain the United States' leadership in space, grow our economy, and strengthen our country's security.

Thank you, Dr. Dittmar, Mr. Hickam, Mr. Kranz, Mr. Stallmer, and Dr. Darden, for joining us today.

In 1961, when President Kennedy announced his ambitious goal for our country to send Americans to the Moon, we did not anticipate the lasting impacts that mission would have on our Nation. At the time, we didn't have the capabilities or know what was needed for mission success.

Apollo 11 showed us what our country and NASA are capable of, demonstrated to the world that the United States is the leader in space, and chartered the space exploration path we continue on today.

The most impressive part is that we developed technologies and prepared for the mission on U.S. soil and my home state of Arizona played a critical role.

The data collected from the Lowell Observatory in Flagstaff, Arizona, was used to make maps of the Moon's surface before the mission. The Apollo astronauts also spent time in Northern Arizona preparing for the mission. They hiked the Grand Canyon to learn about geology and visited Meteor Crater to get an up-close look at what they would encounter on the Moon.

During a test of the first generation space suits at Sunset Crater, also in Arizona, NASA learned that the suits were not thick enough to withstand damage from rocks, forcing a redesign.

The most significant training was done at Cinder Lake, just outside of Flagstaff. Scientists at the U.S. Geological Survey developed a 500-square-foot lunar environment, including 47 craters, to test wheeled rovers, hand tools, and scientific instruments.

These efforts show what we can do as a country when faced with a challenge and success achieving our goal brings.

National and political support, robust funding, and innovation made President Kennedy's goal a reality on July 20, 1969.

As we enter the next phase of space exploration and return to the Moon, we need to continue to utilize American expertise and ingenuity and we need to work closely with our research universities, like the University of Arizona and Arizona State University, that provide leadership and do important work in these areas, and I'm looking forward to holding a hearing on university partnerships this year in this Subcommittee.

The United States has made significant technological advances since 1969 and we have a better idea of what is needed to explore space, but we still face many challenges.

Our workforce is aging and we have not sent humans to space on a NASA spacecraft in 8 years. We must evaluate the best use of taxpayer dollars to achieve our goals and maintain our leadership in space.

Thank you again to all of our witnesses, and I very much look forward to your testimony today.

Thank you, Mr. Chairman. I yield back.

Senator CRUZ. Thank you.

I'd now like to introduce our distinguished panel of witnesses.

Our first witness is Mr. Gene Kranz, who is a retired NASA flight director and fighter pilot. In 1994, after 37 years of legendary Federal service, Mr. Kranz retired from NASA.

After college, Mr. Kranz worked as a flight test engineer for McDonnell Aircraft, developing the Quail Decoy Missile for B–52 and B–47 aircraft. In 1960, Kranz joined the NASA Space Task Group at Langley, Virginia, and was assigned the position of assistant flight director for Project Mercury. He assumed flight director duties for all Project Gemini missions and was branch chief for Flight Control Operations.

He was selected as division chief for Flight Control in 1968 and continued his duties as a flight director for the Apollo 11 Lunar Landing before taking over the leadership of the Apollo 13 Tiger Team.

He was discharged from the Air Force Reserves as a captain in 1972. Mr. Kranz has received many awards and honors, including the Presidential Medal of Freedom, which he received from President Nixon for the Apollo 13 mission, and his designation as a Distinguished Member of the Senior Executive Service by President Reagan.

Mr. Kranz received a Bachelor of Science degree in Aeronautical Engineering from Parks College of St. Louis University.

Our second witness is Dr. Christine Darden. It's good to see you again.

Dr. Darden spent an esteemed 40 years at NASA, becoming one of the world's experts on sonic boom predictions, sonic boom minimization, and supersonic wing design.

During her career, she was appointed as the technical leader of NASA's Sonic Boom Group of the Vehicle Integration Branch of the High-Speed Research Program where she was responsible for developing the Sonic Boom Research Program internally at NASA.

In 1999, she was appointed as the Director in the Program Management Office of the Aerospace Performing Center where she was responsible for Langley Research in Air Traffic Management and Other Aeronautics Programs managed at other NASA centers.

Dr. Darden also served as a technical consultant on numerous government and private projects. She is the author of more than 50 publications in the field of high-lift wing design in super-sonic flow, flap design, sonic boom prediction, and sonic boom minimization.

She earned a Bachelor of Science degree in Mathematics Education from Hampton Institute, a Master's of Science degree in Applied Mathematics from Virginia State College, and a Ph.D. in Mechanical Engineering from George Washington University.

Our third witness is Dr. Mary Lynne Dittmar, who is the President and CEO of the Coalition for Deep Space Exploration.

A 25-year veteran of the space industry, Dr. Dittmar assumed leadership of the Coalition in 2015. Prior to joining the Coalition, from 2012 to 2014, she served as a member of the National Research Council Committee on Human Space Flight.

Prior to that, she acted as a special advisor to the NASA Astronaut Office before her appointment as Boeing Chief Scientist for Commercial Utilization of the ISS.

Dr. Dittmar also coordinated R&D and managed Flight Operations for the Boeing Company on the International Space Station Program.

Dr. Dittmar is a Fellow of the National Research Society and an Associate Fellow of the American Institute for Astronautics and Aeronautics.

Additionally, in June 2018, she was appointed to the Users Advisory Group of the National Space Council. In October of that year was appointed by the Secretary of the Department of Transportation to the Commercial Space Transportation Advisory Committee for the FAA.

Our fourth witness is Mr. Homer Hickam, who is best known for his memoir Rocket Boys about his West Virginia boyhood building model rockets. The book was subsequently made into the film October Sky.

Mr. Hickam is a Vietnam veteran of the 4th Infantry Division and a 30-year careerist with the Army Missile Command and NASA, where he trained astronauts on such missions as Space Lab and the Hubble Space Telescope Repair Mission.

He rounded out his career by negotiating with the Russians on how to train crews on the International Space Station. Besides his career as a writer, Mr. Hickam presently is the Chairman of the Board of the U.S. Space and Rocket Center in Huntsville, Alabama. Mr. Hickam received a Bachelor of Science degree in Industrial Engineering from Virginia Tech.

And our final witness is Mr. Eric Stallmer, who is the President of the Commercial Spaceflight Federation, also known as CSF.

CSF is the largest trade organization dedicated to promoting the development of commercial spaceflight, pursuing ever-higher levels of safety and sharing best practices and expertise throughout the industry.

Before working at CSF, Mr. Stallmer served as Vice President of Government Relations at Analytical Graphics, Inc., AGI. Mr. Stallmer joined AGI in 2002. While there, he oversaw all Washington operations and represented AGI's commercial off-the-shelf products and technology to defense, intelligence, Congress, and civil government sectors within the aerospace industry.

For more than two decades, Mr. Stallmer has also served as an officer in the United States Army and Army Reserves. He is cur rently assigned to the Pentagon and the Office of the Deputy Chief of Staff Army for Logist4ics.

Mr. Stallmer earned a Master's of Arts degree in Public Administration from George Mason University and a Bachelor of Arts degree in Political Science and History from Mount St. Mary College.

Welcome to each of you and, Mr. Kranz, you may begin. Could you turn your mic on, please?

Mr. KRANZ. I had a green button but now it's red.

Senator CRUZ. And I do have——

Mr. KRANZ. Green means go.

Senator CRUZ. And I do have to say I never imagined I would be giving you technical advice.

[Laughter.]

Mr. KRANZ. OK. I'll start over.

Statement of Eugene F. Kranz, Former Apollo Flight Director, Speaker and Author

Mr. KRANZ. Chairman Cruz, Ranking Member Sinema, and Members of the Subcommittee, thank you for the opportunity to discuss NASA's early and future human spaceflight programs.

You know, this is an exciting time for me and for NASA and the space industry as we celebrate one of our Nation's greatest technological achievements, landing two American astronauts on the Moon and returning them safely to Earth on the Apollo 11 mission. I was fortunate to be a part of that team for that endeavor. Growing up, I could have never imagined I would serve in such a role. As a young boy, all I wanted to do was fly. On the path to becoming an aviator, however, I learned of resilience.

I received an appointment to the U.S. Naval Academy, but unfortunately, I failed the entrance physical and I believed my dream was gone.

I obtained a loan and attended Parks College of St. Louis University and earned a degree in aeronautical engineering, and I received an Air Force ROTC Commission.

After tours in Korea as a fighter pilot, I selected Reserve status and in 1958 was assigned as a civilian flight test engineer on the B–52 Program. At the completion of the Flight Test Program, I applied to NASA and was selected to join the Space Task Group at Langley in 1960.

Beginning with John Glenn's Mercury Mission, I served under Christopher Kraft as the assistant flight director. Having never met him before, our initial introduction was short and to the point. He tapped me on the shoulder, said I'm Chris Kraft, you work for me. I want you to go down to the Cape, write some mission rules. When you're ready, give me a call and I'll come down and watch. That was Redstone 1 and I had been two weeks on the job.

As space missions became more challenging and the stakes became higher, we grew into a stronger, more unified and incredibly capable team.

During this period, we developed a set of values. We called them the "foundations," which have guided our operations for the past 50 years and still apply to our work today.

For the past several weeks, I've done dozens of interviews for local, national, and global media, and many reporters have ques tioned whether we should go back to the Moon, should we go to Mars directly and skip the Moon, can we do it again, and why haven't we done it already? All good questions. Should we go back to the Moon? The answer was simply yes, no questions.

There are tremendous opportunities that lunar missions would provide our space industry, includng developing the new capabilities and technologies for space flight missions to the Moon and then beyond.

My answer, can we do it again and why haven't we done it already, is much more complex. That's really why I'm here today, to offer some perspective, based on my experience as a leader of one of the space flight teams which accomplished President Kennedy's 1961 mandate to land an American on the Moon and return him safely to Earth.

The 1960s were not dissimilar to some extent where our Nation is today. President Kennedy faced a confident Soviet Union and a sleeping giant in the People's Republic of China. We were at the beginning of the Vietnam War and the domestic turmoil over civil rights was building.

Getting to Kennedy's goal was timely and masterful in utilizing the challenge of space exploration to unify our Nation and demonstrate the technical capabilities of the United States.

Today, we have many of the same issues. However, one critically important element is still missing. Kennedy's mandate was the impetus but there was a national unity that assured our success.

I believe that today in our country, unity is necessary for great effort and it's lacking within our country, our government, and within the space industry. We have an Administration that is strongly supportive of space, that is willing to provide the resources. We have an agency chartered to do the mission, top level leadership in place, and a very capable work force.

But each of the segments are philosophically divided on the goal. There's infinitely more technological capability than in the early programs, but there is a lack of focus and prioritization.

I believe that the general support for space and a desire to see our Nation explore space will continue but without unity, the space exploration program will be grounded.

To answer the question, what made Apollo successful, it was leadership, unity, and the team. The Mercury and the Gemini Programs provided the knowledge, experience, and environment, and it developed the teams and the technologies and provided the training ground for time-critical, complex, high-risk leadership.

The three elements of the Space Task Group were literally incredible. They created a unique organizational energy of classical aeronautical engineers from the Langley Research Center. They had a team of flight test personnel from Avro, Canada, and we had knowledgeable and energetic young recruits from America's colleges, resulting in a program-wide unity that was focused on a singular objective: space and the Moon.

As the programs evolved, we came face to face with various challenges and failures. We truly began to solidify our team's values and our values were simple: discipline, competence, confidence, responsibility, and team work.

Toughness entered our vocabulary and it was learning the hard way after the Apollo 1 fire and the loss of the crew. Toughness meaning we are forever accountable for actions for what we do or in the case of Apollo 1 what we failed to do.

The leadership inherited from Langley and the Avro group developed organizations with leadership in all segments and every level. There were individuals capable of taking leader-like actions to make their piece work. Leaders with confidence in their ability to send word back up the line that designs, plans, and policies needed amendment or reversal.

Our mission, one team and one voice, was present in every aspect of our work, from the formation of the Space Task Group through subsequent programs.

The three primary elements, however, they contributed to success of the Apollo Program are well documented as NASA's Special Publication 287. They're spacecraft hardware that is the most reliable, flight mission is extremely well planned and executed, and flight crews superbly trained and skilled.

The key that I considered most pertinent, however, to system design is related to safety. NASA has six decades of experience in manned space flight and has written numerous papers related to design criteria, materials, fault tolerance, propellants testing, and many other space system designs.

From the start of Gemini and for subsequent programs, the NASA astronauts, safety engineers, design engineers, and my personnel were embedded in the space system design and change control process from the very beginning of program initiation.

This assured timely inputs to the system design, testing, development of the flight procedures and rules, the plans, and the correct configuration of the mission facilities and trailers.

It is essential in today's card programs that NASA has the insight necessary to ensure safe and successful design, test, and operations. With the emphasis on concurrent engineering and reinvesting NASA, we must assure that individual responsibility is not forever lost. For all of NASA's programs, we must have individuals accountable for design, development, and operations.

While the world has changed dramatically since Apollo and the Space Program since my retirement, the constant essentials for success are unchanged. Leadership, unity, and team work.

I thank you for the opportunity to testify and I look forward to answering your questions.

[The prepared statement of Mr. Kranz follows:]

Prepared Statement of Eugene F. Kranz, Former Apollo Flight Director, Speaker and Author

Chairman Cruz, Ranking Member Sinema, and Members of the Subcommittees, thank you for the opportunity to appear before you today to discuss NASA's early and future human spaceflight programs.

This is an exciting time for NASA and the space industry as we celebrate one of our Nation's greatest technical achievements—landing two American astronauts on the Moon and returning them safely back to Earth on the Apollo XI mission.

I was tremendously fortunate to be a part of the team and that great endeavor. Growing up, I could have never imagined I would serve in such a role. As a young boy, all I wanted to do was fly airplanes—that was my dream. I grew up during the depression, my father was a World War I veteran and died when I was only seven. We lived near the American Legion and, to support her three children, my mother opened a boarding house for military personnel. The influence of these military servicemen really sparked my drive to become a naval aviator.

On the path to becoming an aviator, I learned resilience, very early. I received an appointment to the U.S. Naval Academy but, unfortunately, failed the entrance physical and believed my dream was gone. Other than this appointment, there were no other scholarships pursued and we certainly didn't have the financial means. But an angel in my life, Sister Mary Mark, one of the nuns at Central Catholic High School in Toledo, found an Elks scholarship for Parks College of Saint Louis University. I earned a degree in Aeronautical Engineering and an ROTC commission in 1954. I flew the first three mass produced jet fighters, the F–80, F–86, and the F–100. After a tour in Korea, I selected reserve status and in 1958 was assigned as a civilian Flight Test Engineer on the B–52 at Holloman AFB.

At the completion of the flight test program, I applied to NASA and was selected to join the Space Task Group at Langley in 1960, serving under

Christopher Kraft, as the assistant flight director for John Glenn's Mercury mission. Having never met him before, our initial introduction was short and to the point—he tapped me on the shoulder and stated, "I'm Chris Kraft, you work for me, I want you to go down to the Cape and write a countdown and some mission rules. When you're ready, give me a call, and we'll come down and launch." This was our first Redstone launch. While the space missions became more challenging and the stakes higher, we grew into a stronger, more unified, and capable team. During this period, we developed a set of cultural values which guided our operations over 50 years ago and throughout the decades to come. These steadfast values still apply to our space programs, today.

For the past several weeks, I have done dozens of interviews for local, national, and global media. Besides the typica l questions, "how do you feel about the 50th anniversary and what do you think about the Apollo Mission Control Center restoration?" Nearly all the reporters have followed with questions about our current space initiatives. Should the U.S. go back to the Moon? Should we go to Mars and skip the Moon? Can we do it again and why haven't we already?

My answer to the question, "should we go back to the Moon" is simply—yes. There are tremendous opportunities lunar missions can provide our space industry from redeveloping the capability and honing the spaceflight experience for missions beyond the Moon, as well as commercial development and utilization of the resources. The benefits go well-beyond the time available today and there are plenty of scientists and entrepreneurs who can expound on the economic benefits.

To answer the question, "Can we do it again and why haven't we already," is much more complex. But that's why I am here today, to offer my perspectives based on my experience from having been one of many, key NASA leaders and part of the spaceflight team which accomplished President Kennedy's,1961 mandate, "to land a man on the Moon and return him safely to Earth before the decade was out."

The 1960s were not dissimilar to where our Nation is today relative to geopolitics, domestic turmoil, and patriotism. President Kennedy faced a confident Soviet Union and a sleeping giant in the Peoples Republic of China. We were at the beginning of the Vietnam War and domestic turmoil over civil rights was building. President Kennedy's challenge was a timely, masterful distraction, it utilized the challenge of space exploration to unify our Nation and demonstrated the technical prowess of the United States.

Today, we have many of the same issues. However, unity, was essential to our success in the 60s—one goal, one team.

However, I believe the most critical element we lack today is unity—across our country, our government, and within the space industry.

We have an Administration which is strongly supportive of space, clearly stated directives to align with the goals, and a willingness to provide the resources. We have a capable workforce and industrial base which are engaged in various aspects of the industry, they are anxious to explore, but philosophically divided in their business and technical paradigms. We have infinitely more technological capability than we had in the early programs, but there seems to be a lack of focus and prioritization of those what are explicitly needed to accomplish the mission. We have a Congress divided primarily down partisan lines on just about everything but, individually, have a passion for space and a desire to see our Nation continuing to explore.

It is my understanding that the objective of today's hearing is to address those qualities that were contributors to the success of the Apollo Program. Many papers have been written on this topic so my comments should not be new.

What made Apollo a success? The Leadership. The Unity. The Team

The Mercury and Gemini Programs provided the experience base for the Apollo, Shuttle, and subsequent programs.

The Mercury Program provided the initial understanding of the manned flight mission environment, the involved mission tasks, facility requirements and capabilities, training and the nature of the personnel most suited for the critical, high risk aspects of mission control.

The Gemini Program introduced the critical new space technologies; computers, fuel cells, maneuvering and attitude control systems, and the ability to accomplish rendezvous, docking, and support extravehicular operations. Computers, satellite communications, and improvements in display technology in Mission Control combined with effective simulation training established the relationship between the mission control team and crew needed for complex space operations. Possibly the most important was that it provided the training ground and mission experience required for making risk-based decisions during premission and mission operations.

Developing our Team and Our Values

The Space Task Group was an enterprise, with three unique components. The foundation was provided by a small group of classical aeronautical engineers from the Langley Research Center. This group had the hands-on knowledge and experience from their work in the design and test of the breakthrough aircraft during and after World War II. The second group was composed of aeronautical engineers and flight test personnel from the Avro Arrow project in Canada. The cancellation of the AVRO Arrow project by the Canadian government made key engineers and flight test personnel available to the fledgling Space Task Group. The third component was comprised of a mixture of young engineers recruited from America's colleges, former military personnel, and a small group experienced in early scientific satellite programs.

Each of these groups brought with them a unique organizational chemistry. The Langley group brought a classical aeronautical engineering skill to the Mercury Program. The Arrow group brought an incremental flight test approach. The third, younger group, brought a highly energetic and impatient, "let's get going" approach. The interaction of these three groups created an organizational chemistry that was greater than the sum of its parts that I believe led to developing the incredible and gifted leadership that provided success during the early programs.

As the programs evolved and we came face-to-face with various challenges and failures, we truly began to solidify our team core values of discipline, competence, confidence, responsibility, and teamwork (ref. Foundations of Mission Control). Toughness, emerged as a core value, learned the hard way, after the Apollo 1 fire and the loss of the crew. Toughness, meaning, we are forever accountable for our actions, what we do, or, in the case of Apollo 1, what we failed to do (ref. Kranz Dictum).

The organizational structure for future space programs must consider the relationship between Headquarters personnel and those assigned to mission leadership. From Gemini through the early Shuttle Program a highly professional, personal, and friendly relationship existed to address issues on a variety of issues before they became problems.

Leadership

The leadership inherited from the Langley and Arrow groups recognized that for an organization to function, leadership must exist in all segments and at every level. There must be individuals capable of taking leader-like actions to make their piece work, leaders with confidence in their ability to send word back up the line that design, plans and policies needed amendment or reversal. There was a universal recognition that every member of the Space Task Group was responsible to develop the next generation of leaders. During the Shuttle Program, many organizational "fads" originated. Awards were given for "flattening organizations," essentially eliminating mid-level supervisors. I believe this was detrimental, mid-level positions provide the primary training ground for higher level positions.

Teamwork and Unity

One mission, one team, one voice was present in every aspect of our work from the formation of the Space Task Group through all subsequent programs. There was so much to be learned, and work to be done that unification of both NASA and contractor organizations in every activity was universally recognized as essential to the program success. The unity I had seen through the early Shuttle Program began to shatter, reaching crescendo on the Space Station Program in the period after the Space Shuttle Challenger accident resulting in lost opportunities, schedule and cost impacts, and many good leaders deciding it was time to retire.

Policy

The three basic elements that contributed to the success of the Apollo Program are: spacecraft hardware that is the most reliable, flight missions that are extremely well planned and executed, and flight crews that are superbly trained and skilled. These policies have guided me and my teams in all programs. NASA document SP– 287, What Made Apollo a Success, provides many of the specifics related to the Apollo Program. This report addresses spacecraft development, mission development, flexible yet disciplined mission planning and execution, flight crew training, and trajectory control techniques.

A key area I consider pertinent to system design is related to safety. NASA has six decades of experience in manned space flight, has written numerous papers related to design criteria, materials, fault tolerance, propellants, testing, and many other space systems design elements.

From the start of Gemini and for subsequent programs, the NASA astronauts, safety engineers, design engineers, and operational personal were embedded in the space systems design and change control process at program initiation. I had direct communications with the prime contractors' design and test organizations. This assured timely operational inputs to the space system design, development of flight procedures and plans, and the configuration of mission facilities and trainers. My controllers were some of the best systems engineers in the world. Astronaut John Young assured the mission control cadre participated in every major accident review and contributed to the redesign when needed.

In today's "Commercial Crew and Commercial Cargo," NASA gets to see what is designed after the fact and often too late to make any critical changes." I have asked NASA personnel on several occasions who is accountable for providing oversight to the commercial crew space systems design, test, and operations. The answers I received were not specific.

The day I retired, I left a memo for Dave Leestma, Chief of the Astronaut Office, containing testimony by Admiral Rickover on accountability.

The Rickover memo described the lack of accountability for the 1963 submarine Thresher loss during a diving test with all crew aboard. This was the environment I experienced in the post-Challenger and early Space Station period.

> "During the six years of the submarines design, the Portsmouth Naval shipyard had three shipyard commanders, three production officers and five planning officers.
>
> The Bureau of Ships during this period had two Chiefs of Bureau, five or six chiefs of the design division and three heads of the Submarine Type desk."

I closed my memo to Leestma with these words, "With the emphasis on concurrent engineering and reinventing NASA, we must assure that individual responsibility is not forever lost. When the dust finally settles on the trials and tribulations of our programs, we must have individuals accountable for design, development, and operations."

If a crew is lost on a commercial crew mission, who will be held accountable?

While the world has changed dramatically since Apollo, and in the Space Program since my retirement in 1994, the one constant essential to success, is unchanged, it is leadership.

John Gardner, in the preference to his excellent book "On Leadership" states, "In order for an organization to function, leadership must be dispersed throughout all segments, and at every level there must be individuals capable of taking leader like actions to make their piece work. Men and women who are not afraid to send word back up the line that newly announced plans and policies need amendment or reversal."

> Gardner then comments on the large numbers of people who are torn loose from values they may have held previously, what he calls the divergence of value systems. "Leaders are always seeking the common ground that will make concerted action possible. *It is impossible to exercise leadership if shared values have disinte-grated."*

In conclusion, I believe the book, "Apollo the Race to the Moon" by Charles Murray and Catherine Bly Cox, provides an in-depth perspective of the programmatic, engineering, and operational elements responsible for the success of the Apollo Program. I would recommend that this book is made "required reading" for those who would assume future leadership and programmatic functions.

Today, our National leadership and the NASA industry team are at a critical, "go, no-go" point. "Now is the time to take longer strides . . . time for this Nation to take a clearly leading role in space achievement which, in many ways, may hold the key to our future on Earth." President John F. Kennedy.

Thank you for the opportunity to testify. I look forward to answering your questions.

Foundations of Mission Control

- **To instill within ourselves these qualities essential to professional excellence:**

  Discipline – Being able to follow as well as to lead, knowing that we must master ourselves before we can master our task.

  Competence – There being no substitute for total preparation and complete dedication, for space will not tolerate the careless or indifferent.

  Confidence – Believing in ourselves as well as others, knowing that we must master fear and hesitation before we can succeed.

  Responsibility – Realizing that it cannot be shifted to others, for it belongs to each of us; we must answer for what we do, or fail to do.

  Toughness – Taking a stand when we must; to try again, and again, even if it means following a more difficult path.

  Teamwork – Respecting and utilizing the abilities of others, realizing that we work toward a common goal, for success depends upon the efforts of all.

  Vigilance – Always attentive to the dangers of spaceflight; Never accepting success as a substitute for rigor in everything we do.

- **To always be aware that suddenly and unexpectedly we may find ourselves in a role where our performance has ultimate consequences.**

- **To recognize that the greatest error is not to have tried and failed, but that in the trying we do not give it our best effort.**

The Kranz Dictum

(speech to the control team after the Apollo I fire)

Spaceflight will never tolerate carelessness, incapacity, and neglect. Somewhere, somehow, we screwed up. It could have been in design, build, or test. Whatever it was, we should have caught it. We were too gung-ho about the schedule and we locked out all of the problems we saw each day in our work. Every element of the program was in trouble and so were we. The simulators were not working, Mission Control was behind in virtually every area, and the flight and test procedures changed daily. Nothing we did had any shelf life. Not one of us stood up and said, 'Dammit, stop!' I don't know what Thompson's committee will find as the cause, but I know what I find. We are the cause! We were not ready! We did not do our job. We were rolling the dice, hoping that things would come together by launch day, when in our hearts we knew it would take a miracle. We were pushing the schedule and betting that the Cape would slip before we did.

From this day forward, Flight Control will be known by two words: 'Tough' and 'Competent.' Tough means we are forever accountable for what we do or what we fail to do. We will never again compromise our responsibilities. Every time we walk into Mission Control we will know what we stand for. Competent means we will never take anything for granted. We will never be found short in our knowledge and in our skills. Mission Control will be perfect. When you leave this meeting today you will go to your office and the first thing you will do there is to write 'Tough and Competent' on your blackboards. It will never be erased. Each day when you enter the room these words will remind you of the price paid by Grissom, White, and Chaffee. These words are the price of admission to the ranks of Mission Control.

Senator CRUZ. Thank you, Mr. Kranz. Dr. Darden.

Statement of Dr. Christine M. Darden, Retired, NASA Langley Research Center

Dr. DARDEN. Mr. Chairman, Members of the Senate Commerce Committee, and Members of the Subcommittee on Aviation and Space, I am very honored to be present today to be participating in your committee hearings.

I am here as a child of the Space Race. On October 5, 1957, I vividly recall the headline about Sputnik as I took the newspaper into my high school library.

I also vividly remember May 1961 when my dormitory halls were filled with screams about the challenge that President John Kennedy had just issued about landing a man on the Moon and safely bringing him back before the end of the decade.

Where have we been? It is the response to that challenge with Apollo that makes NASA and all of us so very proud. At that time, NASA Langley Research Center had existed as the National Advisory Committee for Aeronautics, NACA, Laboratory for 40 years and had spent countless hours thinking about how to leverage our aeronautics expertise to address the challenges of spaceflight.

NASA Langley had begun Project Mercury and trained its seven Mercury astronauts to fly in space. As preparation for Apollo, the Gemini Program focused on rendezvous, docking, and long-term space flight.

The Apollo Program captured the world's attention and demonstrated the power of America's vision and technology to inspire great achievements.

On July 16, 1969, though it was the NASA Space Centers that were visible to the public as the Saturn 5 set poised for liftoff, many workers from the Research Centers, like Langley, were several years removed from the simulators that they built for training the astronauts, from the wind tunnel tests that they did, which is where I worked. Those tests validated the spacecraft configuration and determined the re-entry safety.

But Apollo was in their hearts, and Apollo was supported by this country. After 11 and a half years, Apollo ended, having spent a record $23 and a half billion, having placed on the Moon and safely returned 12 men, and having inspired thousands of young engineers and space enthusiasts, like myself.

Sadly, after Apollo ended, there was a decline in the number of American students getting higher degrees in the STEM areas.

Where do we go from here? NASA is now poised to return American astronauts to the lunar surface by 2024 as part of our broader Moon-to-Mars Exploration. Artemis is the program to prove technologies, capabilities, and new business approaches for future missions to Mars. Through Artemis, NASA will establish a sustainable human presence on the Moon by 2028, inspiring the Artemis generation.

Certainly, one of the necessary needs as we go forward is the development and preparation of a strong, energetic, and committed workforce as we had for Mercury, Gemini, and Apollo. A thriving, visible Artemis Program will do much to inspire the next generation and to pursue STEM careers, just as I was inspired as I saw the work being done at NASA result in Neil Armstrong walking on and safely returning from the Moon.

After Apollo, I spent 25 years working in supersonic aero-dynamics and minimizing the sonic boom. When I retired from NASA, two young ladies came to my party and stated that it was because of my sharing my story of how I got to NASA and what I did at NASA that inspired them to become engineers at NASA. I have over the past two years spoken to thousands of students about following their dream, about preparing themselves, and persevering, hoping that they, too, will be inspired to join the likes of an Apollo generation to push the boundaries of knowledge. They will pursue groundbreaking research to understand how to live and work on another planet.

Since retiring from NASA, I have remained an advocate for the agency and I will work in aeronautics and exploration.

I wish to thank the Committee members for your continued support of NASA and for the important work you are doing to pass NASA's Authorization bill this Congress.

NASA really needs an authorization bill that supports our Nation's plans for the Artemis Program and the overall approach for the Moon and Mars.

In summary, this lunar destination is promoting sustainability, a proving ground for Mars, a strategic presence for our nation, and a foundation for building international and commercial partnerships while also inspiring the next generation to be prepared for the excitement and groundbreaking new opportunities.

I thank you very much.

[The prepared statement of Dr. Darden follows:]

Prepared Statement of Dr. Christine M. Darden, Retired, NASA Langley Research Center

Mr. Chairman, Members of the Senate Commerce Committee, and Members of the Subcommittee on Aviation and Space, I am very honored to be present today to participate in your Committee Hearing. I am here as a child of the Space Race. I remember May 22, 1961, when President John Kennedy issued the challenge that we would put a man on the moon and safely bring him back before the end of the decade.

Where We Have Been

It is the response to that challenge with Apollo that makes NASA and all of us so very proud. At that time, NASA Langley Research Center had existed as the National Advisory Committee for Aeronautics (NACA) for 40 years, and had spent countless hours thinking about how to leverage our aeronautics expertise to address the challenges of spaceflight. NASA Langley had begun Project Mercury and trained its seven Mercury Astronauts to fly in space. The Gemini Program focused on rendezvous, docking and long-term space flight.

The Apollo Program captured the world's attention and demonstrated the power of America's vision and technology to inspire great achievements. On July 16, 1969, though it was the NASA Space Centers that were visible to the public as the Saturn V set poised for lift off, many workers from the Research Centers were now several years removed from the simulators they built for training the astronauts, the wind tunnel tests that validated the configuration and re-entry safety, and the discussions that led to decisions on how we would get to the moon. But Apollo was in their hearts and Apollo was supported by this Country. After 11.5 years Apollo ended, having spent a record $23.5B, having placed on the moon and safely returned 12 men, and having inspired thousands of young engineers and space enthusiasts.

Where We Go from Here!

NASA is now poised to return to the Moon in the next 5 years. This time the plan is to return to the moon as a pathway to further exploration but not alone, rather with government, industry and international partners in a global effort to build and test the systems needed.

Artemis is the program to prove technologies, capabilities and new business approaches for future missions to Mars. NASA has begun the next era of exploration and discovery! The plan is to return American astronauts to the lunar surface by 2024 as part of our broader Moon to Mars exploration approach. The first American woman and next American man will set foot on and explore the South Pole of the Moon, where no human has ever been before.

The first Artemis flight will be a test of the Space Launch System (SLS) rocket and Orion spacecraft as an integrated system. The second Artemis flight will ferry the crew to the Moon aboard SLS and Orion. The third Artemis flight will deliver the crew to the lunar surface. Through Artemis, NASA will establish a sustainable human presence on the Moon by 2028.

Inspiring the Artemis Generation!

A thriving, visible Artemis Program will do much to inspire the next generation to pursue STEM careers. This lunar effort will engage the entire nation and the world—uniting the brightest minds of academia, industry and communities of all sizes and types, from early career professionals to our international partners. Just think about it. The Artemis generation will push the boundaries of human knowledge. They will dream about and eventually pave the way for reaching new worlds and unlocking the mysteries of the universe. They will pursue groundbreaking research to understand how to live and work on another planet. This research will expand our knowledge of human anatomy, solar propulsion, biofuels, geology, astrophysics, and lunar and planetary science.

Since retiring from NASA, I have remained an advocate for the Agency and our work in aeronautics and exploration. I wish to thank the Committee Members for your continued support of NASA and for the important work you are doing to pass a NASA Authorization bill this Congress. NASA really needs an Authorization Bill that supports our Nation's plans for the Artemis Program and the overall approach for Moon to Mars.

In summary, this lunar destination is promoting sustainability, a proving ground for Mars, a strategic presence for our nation, and a foundation for building international and commercial partnerships while also inspiring the next generation to be prepared for the excitement of new opportunities.

Thank you.

Senator CRUZ. Thank you, Dr. Darden. Dr. Dittmar.

Statement of Dr. Mary Lynne Dittmar, President and Chief Executive Officer, Coalition for Deep Space Exploration

Dr. DITTMAR. Chairman Cruz, Ranking Member Sinema, Members of the Subcommittee, thank you for the invitation to appear before you today with such an extraordinary and distinguished panel to discuss my thoughts on the topic of today's hearing.

More than 50 years ago, the Apollo Program began, ultimately resulting in one of humankind's most extraordinary achievements. Mission readiness was the driving force of the entire effort.

To achieve the mission, certain capabilities had to be developed. These included standing up a human space flight organization capable of developing and delivering a super-heavy lift vehicle, a crew capsule, ground systems, AVA suits, cruise systems, silence payloads, and related equipment. An entire operations concept and an organization called Mission Control had to be invented. I will defer here to Mr. Kranz.

The overarching goal, however, was the geopolitical one. The Apollo Program was the means to achieve it. Involving over 400,000 Americans and at the cost of 300 billion in adjusted dollars for the entire program, President Kennedy's challenge was fulfilled between July 16 and 24, 1969, with the lunar landing on July 20, 50 years ago.

With regard to today's goals, the primary objective of NASA's Human Space Flight Program viewed through a national lens remains a geopolitical one, although it is a different geopolitical objective. No longer in a race to the Moon with the Soviets, United States leadership in space depends upon creating a foundation that provides other nations and a nascent space-based economy with security and assurance regarding our national intentions and long- term commitment to aspire, inspire, and achieve, in short, to lead human space exploration and the development of space.

If we do not do this, rest assured that someone else will. Space remains the strategic competitive domain between nations.

Accordingly, the United States is developing core capabilities to enable our return to deep space. A super-heavy launch vehicle, a modern crew vehicle capable of long duration missions when paired with habitats and consumables, and ground-based infrastructure needed to support those missions.

Dubbed the Space Launch System, the Orion Crew Vehicle, and Exploration Ground Systems, respectively, these strategic assets are the foundation upon which national goals and human deep space exploration will rest for the foreseeable future.

Similar to the development of military capabilities, these are long lead time national assets, sustained as a guarantee against economic downturn, policy shifts, and as a message to the global community.

In addition, NASA's Artemis Program will engage industry and international partners. Sticking to aligned national objectives with those of commercial enterprises and global collaborators in the human exploration of deep space.

The momentum established by the current Administration with its focus on the Moon is a welcome one, but as report after report has shown, NASA is asked to do too much with too little and this is no exception.

Acceleration must be balanced with management of program risk, mission assurance, and safety. Forward momentum also must be matched by significant national investment above present levels. The Administration has proposed $1.6 billion to begin the process of accelerating our Human Space Exploration Program. In addition, the NASA Administrator has estimated it will take between four to six billion per year over currently anticipated funding levels to meet more aggressive timelines.

While funding increases are always a political challenge, it is worth noting that the benefits of 10 times that amount in adjusted dollars invested in the Apollo Program are evident to all and form the foundation both for today's national effort and for the growing entrepreneurial sector.

A personal aside, I actually have been on crutches for the last two and a half years as a result of a connective tissue disorder, and during the launch and landing of Apollo 11, I was at home after having just gone through my fourth surgery. I now stand before you without crutches and due to two years of work in physical therapy using technology that was developed in the Space Program. So for me, this is deeply personal.

NASA plans to achieve architectural flexibility at the Moon by constructing a lunar orbital station, known as Gateway, in partnership with industry and international partners.

Among other things, Gateway simplifies our ability to aggregate hardware in the vicinity of the Moon, including descent and ascent vehicles for transportation of humans back and forth to the lunar surface.

Longer surface missions enabled by Gateway make it easier to conduct lunar operations over time. Entrepreneurial firms are teaming with investors and also established companies to develop technologies aimed at building up infrastructure on the Moon.

With regard to science, the Moon has been called the Rosetta stone of the Solar System with evidence locked within it that has already taught us a great deal about the formation of the Earth-Moon system. We've explored only 5 percent of the surface. There's a great deal more to learn.

Finally, Gateway is the next logical step in developing a command and logistics capability that is extensible not just to the Moon but beyond the Moon toward Mars.

The Gateway itself is a prototype, an evolution of lessons learned over the last 50 years and, in particular, from the International Space Station to create habitable space and systems to support human life and work in deep space.

The Moon is an important stepping stone. What we learn there will not only create opportunities in lunar space but will open new discoveries and knowledge that would help us as we look toward Mars. But the Moon is not an end goal, it's the beginning, a next step enabling the migration of technology, heavy industry, and humanity itself off the Earth and into the Solar System at a scale that is no longer constrained by a single planet, our original home.

Thank you for your time and attention. I look forward to any questions you may have.

[The prepared statement of Dr. Dittmar follows:]

Prepared Statement of Dr. Mary Lynne Dittmar, President and Chief Executive Officer, Coalition for Deep Space Exploration

Chairman Cruz, Ranking Member Sinema, Members of the Subcommittee, thank you for the invitation to appear before you with such an extraordinary and distinguished panel to discuss my thoughts on the topic of today's hearing: "NASA's Exploration Plans—Where We've Been and Where We're Going."

Then: The Apollo Program and an Epochal Mission

More than 50 years ago, the Apollo program began, ultimately resulting in one of humankind's most extraordinary achievements. Undertaken as part of the Kennedy administration response to the Soviet Union, and in the context of a perceived "missile gap" between U.S. and Soviet capabilities, the effort was virtually on a war footing and had a specific mission focus— "landing a man on the moon and returning him safely to the Earth" within a decade. This mission objective had a much broader and even more urgent underlying goal, however—to demonstrate the superiority of the technical, economic, and political system represented by American ideals of freedom and democracy versus that of a closed, communist state.

The mission outlined by Kennedy may not have survived had he not been assassinated—and if it had survived it may have evolved to a different mission, perhaps with a Russian partner. We will never know. In President Johnson's hands, however, Kennedy's space legacy—both the goal and the mission—became codified in a way that has never happened since in human space exploration.

Mission readiness was the driving force for the entire Apollo program. To achieve the mission certain capabilities had to be developed. These included standing up a human spaceflight organization capable of developing and delivering a super heavy lift vehicle, a crew capsule, ground systems, EVA suits, crew systems, and related equipment. An entire operations concept and organization called "Mission Control" had to be invented, in turn supporting and supported by development and execution of procedures, schedules, simulations, training, a logistics system (including communications and tracking both on the ground and from ground to space and back), medical support, and much more. Each capability was developed as part of a learning sequence with many moving parts, each building toward the successful landing and safe return of the crew of Apollo 11. The goal was a geopolitical one, the Apollo program was the means to achieve it, and the extraordinary work, time, and treasure it took to meet it—involving over 400,000 Americans and at the cost of $300B in adjusted dollars for the entire program[1]—was realized between July 16 and 24, 1969, with human beings setting foot on another celestial body 50 years ago on July 20.

Now: A Focus on Capability

With regard to today's goals, it should be said from the beginning that the primary objective viewed with a national lens remains geopolitical, although it is a different geopolitical objective. No longer in a race to the Moon, United States leadership in space depends upon establishing a foundation that provides other nations and a nascent space-based economy with security and assurance regarding our national intentions and long-term commitment to aspire, inspire and achieve—in short, to lead. Space remains a strategic, competitive domain among nations. For example, China is pushing forward with technology, science, investment, a new space station in low Earth orbit, and lunar aspirations, and has announced development of a super heavy lift

[1] http://www.planetary.org/press-room/releases/2019/apollo-cost-analysis.html.

vehicle (SHLV) for human space exploration. Russia remains an active partner with us on the International Space Station but also has announced plans for an SHLV. For the U.S. to push forward and support our endeavors and those of our friends, we must be "out there"—physically present, with national assets at the ready, and we must be there sooner rather than later.

The success of the Apollo program—at great cost in treasure and in human lives— created a near-mythical template in the minds of both NASA and the public that has been at the heart of controversy about the role of the space program ever since, resulting in oft-repeated commentary as to whether NASA has a "mission." The answer is emphatically "yes," but not in the same way as Apollo. Instead, we are embarked upon a plan to create, manage, and execute deep space activities based on the development of national capability, defined here as the ability to achieve a variety of desired outcomes in a specific operating environment. Much like the development of military capability, a sustained national capability requires technical systems and equipment needed to perform the operations for which they are designed that also support a variety of missions that may not be known when the capability is being developed. These are typically long lead-time national assets that exist to perform certain functions necessary to meet operational requirements. As such they are sustained by national investment as a guarantee against economic downturns and policy shifts that may accompany short-term Administration and/or Congressional priorities.

Accordingly, the United States is developing core capabilities to enable our return to deep space: A super heavy launch vehicle (SHLV), a modern crew vehicle capable of long-duration missions when paired with habitats and consumables, and ground-based infrastructure needed to support those missions. Dubbed the Space Launch System (SLS), the Orion crew vehicle, and Exploration Ground Systems respectively, these strategic assets are the foundation upon which national goals in human deep space exploration will rest for the foreseeable future. In addition, the Artemis Program will engage industry and international partners, seeking to align national objectives with those of commercial enterprises and global collaboration in the exploration of deep space wherever practicable.[2]

The momentum established by the current administration with its focus on the Moon is a welcome one, but as report after report has shown NASA is asked to do too much with too little and this is no exception. Acceleration must

---

[2] https://www.nasa.gov/sites/default/files/atoms/files/ger_2018_small_mobile.pdf.

be balanced with program risk and mission assurance, and forward momentum must be matched by significant national investment above present levels.

Happily, this increased investment will find a much broader portfolio in science, technology, and exploration than when the Apollo program began 60 years ago. A revitalized space industry is building on previous government investments and experience gained during that extraordinary effort. New manufacturing methods, technologies, and advanced computing capabilities are reducing costs, encouraging new entrants, building new capacity, and attracting billions of dollars in investment in the emerging space sector. Meanwhile, NASA has led the development of a global spaceflight community, collaborating with international partners while helping to open the door to human spaceflight and space science for over 100 countries via participation on the International Space Station.

Today's need for operational readiness is every bit as great as during the Apollo program, but the overarching goal is less specific, more open-ended. While the recently-announced Artemis program, with its goal of returning Americans to the lunar surface by 2024 is a worthy effort, the "Artemis Lunar Landing-1" mission does not represent fulfillment of a policy objective in the same sense as did Apollo 11. Today, the policy objective is to establish a sustainable, strategic human presence beyond low Earth orbit, first at the Moon and then beyond, with an eye toward Mars. The key words here are "sustainable" and "strategic." To succeed in this endeavor, however, several things must change.

Timely and Sufficient Funding In the decades since the current, regular order for authorizing and appropriating tax dollars has been established (1974), Congress has managed to pass all of its required appropriations measures on time on less than half a dozen occasions.[3] It is true that Congressional support for NASA and for human space exploration has been consistent. That said, for NASA programs, as for all long-lead time programs spanning years, last-minute political maneuvering, government shutdowns, delayed receipt of funds, and general instability and unpredictability of funding invariably creates inefficiency in program management and execution, forcing compromises in program planning that add cost, schedule and risk.[4] In addition, report after report has found that NASA's tasking and

---

[3] https://www.pewresearch.org/fact-tank/2018/01/16/congress-has-long-struggled-to-pass- spending-bills-on-time/.

[4] https://www.hsdl.org/?view&did=794133.

budget are mismatched.[5] Alternate acquisition and procurement approaches when technologies are well understood is a useful means to reduce costs, but relying on industry to drive down costs while at the same time increasing speed of delivery, or seizing upon alternate means for acquisition without consideration of risk management implications, amounts to wishful thinking. Such approaches do nothing to change that fact that at no time since Apollo, with all of its extraordinary resources, has any human-rated system been deployed less than 4 years later than originally intended, no matter who is building it.

The Administration has proposed $1.6B to begin the process of accelerating our human space exploration program. In addition, the NASA Administrator has estimated it will take between $20 and $30B over currently anticipated funding levels to meet more aggressive timelines. While funding increases are always a political challenge, it is worth noting that the benefits of 10X that amount in adjusted dollars invested in the Apollo program are evident to all, and form the foundation both for today's national effort and for the growing entrepreneurial sector.

It is absolutely true that that inefficiencies in development and execution within government programs should be addressed. Congress can help by regularizing its own process. The hearing to which these remarks is addressed is a welcome aspect of that process.

Acquisition and Procurement Reform As has been recognized by Congress over the past many years, the nature of the policy and regulatory framework guiding exploration and development of space is crucial, providing government, industry and the investment community alike with stable, predictable operating environments. The importance of stability also has been recognized across various Administrations and very recently has been reflected in the agreement upon 21 "Guidelines on the Long-Term Sustainability of Outer Space Activities" within the UN's Committee on the Peaceful Uses of Outer Space (UN COPUOS).[6]

Hand-in-hand with stability in "light touch" regulation however is the need for streamlined acquisition processes. As I testified before the National Space Council in February of 2018, the fiscal, programmatic, reporting and management burdens associated with government contracting and oversight cannot be overstated. The slow nature of acquisition and the costs and schedule

---

[5] See, for example: https://sites.nationalacademies.org/cs/groups/depssite/documents/webpage/deps_080254.pdf.

[6] http://www.unoosa.org/oosa/en/ourwork/topics/long-term-sustainability-of-outer-space-activities.html.

associated with program start- up also pose a threat to technology development and insertion into some of our most strategic programs, threatening U.S. leadership and security. This is not to say that traditional contracting should be abandoned. On the contrary, cost-plus contracts for managing large development programs with significant R&D components provide stability, reduce business risk and support the aerospace and defense industrial base that is so critical to U.S. security. That said, the ability to ramp up programs with much greater speed than typically seen together with renewed focus on accountability and constraints on cost and schedule growth are of primary interest in reform.[7]

The use of Other Transactional Authority (OTAs) and Public Private Partnerships (P3) may provide more flexibility than traditional contracting and, in some cases, offer economic incentives, but at the cost of reduced transparency. Each of these approaches may take many forms but their success depends upon informed allocation of risk on each side of the government-business relationship, a realistic business case in the case of P3 (an indispensable requirement) and the means to align business objectives with the public (national) interest. The rapidly-accelerating range of procurement activities in both the DoD and NASA requires the ability to evaluate acquisition models, understand their strengths and weaknesses for differing applications, situations and goals, assess business cases, grasp the complexities of risk allocation and management associated with different models, and understand the range of economic incentives and their likely downstream effects. Adopting an acquisition posture that balances risk, cost and schedule with the goal of rapid development and deployment of capabilities may require retraining the procurement workforce across the government.

Education and Diversity in the Workforce

Preparation and development of the A&D workforce is not limited to procurement professionals. It must begin much earlier and span a multitude of disciplines. The United States has remained in the upper third of countries for 4–12 grades students in math and science for the past 15 years though our test

[7] http://exploredeepspace.com/multimedia/coalition-statements/2018-02-22/.

scores remain relatively flat.[8] Of those students who eventually seek employment in the Aerospace and Defense (A&D) workforce, 71 percent of young A&D professionals report they first became interested in these careers during their grade school years. Continuing emphasis upon superior preparation and broad educational opportunities is critical to U.S. leadership across STEM disciplines, including in space.

In addition, and despite significant effort, the diversity of the A&D industry looks much the same as it did four decades ago. Greater diversity in a workforce has been demonstrated time and time again to result in higher rates of innovation, invention, and unique problem-solving and is key to both global competition and cooperation. At the same time, the shift in economic demographics nationwide since 2008 has resulted in a higher proportion of Americans remaining at work. This is also true of A&D; 29.8 percent of workers are 55 years of age or older.[9] On the one hand, this benefits the industry since it requires legacy skills and experience in aerospace, electrical, mechanical and electrical engineers. On the other, the industry also requires skills in cognitive computing, artificial intelligence and materials. At present, terrestrial high technology careers in these fields are attracting the best and brightest young professionals. NASA and the Department of Defense must not only compete with them but with quicker-moving entrepreneurial firms.[10]

While job competition is good to spur interest in STEM careers and to drive up salaries, from the point of view of national interest, the stresses on the workforce (and the hundreds of small businesses that provide the A&D backbone in this country) are increasing, creating "talent gaps" in some critical fields ranging from IT to manufacturing talent—all of which are needed in space. Reorienting and revitalizing human capital processes, increasing investment in trade schools, and even embedding space into higher education curriculums in non-traditional fields are among the means to help address the growing workforce development issue.

---

8 https://www.brookings.edu/blog/brown-center-chalkboard/2017/04/07/what-international-test-scores-reveal-about-american-education/.

9 https://www.aia-aerospace.org/wp-content/uploads/2018/05/AviationWeek-WhitePaper-REVISED-101717.pdf.

10 https://spacenews.com/op-ed-talent-gap-jeopardizes-space-business-national-security/.

The Near Future: Artemis and the Moon

Addressing funding, acquisition, and workforce development and diversity are fundamental to get to where we are going—which is to the Moon and then to Mars. As pointed out by the Committee on Human Space Exploration of the National Academies, upon which I served, there are many roads the United States and its industry and international partners may take to create a sustainable presence beyond low Earth orbit. The "Pathways to Exploration: Rationales and Approaches for a U.S. Program of Human Space Exploration" report (herein after referred to as "The Pathways Report"),[11] published in 2014 after two years of study and debate, described several sequences and risk postures for NASA's human exploration program but leaned heavily toward the Moon as an initial destination. The Pathways report also advised Congress and NASA to focus on "feed forward" development— that is, to maximize the potential for a sustainable approach by designing in readiness and flexibility. Among other means to achieve this design philosophy is to develop capabilities that can be deployed over decades.

Space Policy Directive-1, signed by President Trump in late 2017, is in alignment with the NASA Authorization Acts of 2008, 2010, and 2017 in its call for NASA to "lead an innovative and sustainable program of exploration with commercial and international partners to enable human expansion across the solar system and to bring back to Earth new knowledge and opportunities."[12] While the Administration's target date of 2024 for the Artemis Lunar Landing-1 mission has injected some much-needed urgency into NASA's planning and programs, one danger of shifting our focus from developing a longer-term capability to focus on a near term date is that we may inadvertently sacrifice flexibility and capability for the sake of meeting a short-term mission objective.

In military science, a major area of endeavor is to find methods to achieve missions that may not have been understood when systems were under development, using those capabilities that have since come on line both as intended and by exercising new concepts. With regard to deep space exploration, NASA plans to achieve flexibility at the moon by constructing a lunar orbiting station known as Gateway in partnership with industry and with international partners. Gateway is not, as some have stated, a waste of money

---

[11] https://www.nap.edu/catalog/18801/pathways-to-exploration-rationales-and-approaches- for-a-us-program.

[12] https://www.whitehouse.gov/presidential-actions/presidential-memorandum-reinvigorating-americas-human-space-exploration-program/.

or an unnecessary way station. Instead, Gateway is a component of a larger architecture that increases feed forward opportunities. The Gateway greatly simplifies our ability to aggregate hardware in the vicinity of the Moon—that is, to provide a central hub for docking, brief habitation, emplacement of consumables, descent and ascent vehicles for transportation to the lunar surface, transfer vehicles, etc. It is true that these can be aggregated in orbit without a Gateway; however, the transportation model becomes more complex and the timing less forgiving.

For example, without a Gateway, lunar sorties with humans are limited to four days due to the 21-day mission profile for the Orion crew vehicle. With the Gateway, crew are not solely dependent upon Orion's systems but can rely on habitation and access to consumables that are present at the Gateway, enabling longer missions on the surface. In addition, over time the Gateway will enable in-space reusability by serving as a docking and refueling hub, greatly simplifying transportation requirements over time.

Longer surface missions in turn will enable a variety of activities to be undertaken by a variety of participants. Entrepreneurial firms are teaming with investors and also with established companies to develop technologies aimed at building up infrastructure on the Moon, making use of lunar regolith as raw material as one example. With regard to science, there is still a great deal to learn. The Moon has been called "The Rosetta Stone"[13] of the solar system, with evidence locked within that has already taught us a great deal about the formation of the Earth-Moon system but also has implications for outer bodies. We have explored only 5 percent of the surface; there is a great deal more to learn.

Finally, Gateway is the next logical step in developing a command and logistics capability that is extensible not just to the Moon but beyond the Moon toward Mars. The Gateway itself is a prototype, and evolution of lessons learned over the past 50 years and in particular from the International Space Station to create habitable space and systems to support human life and work in deep space. As the Gateway evolves, it provides an opportunity for broader international and industry participation and utilization, providing options for continuing deep space development that do not necessarily depend upon descending to the surface.

---

[13] https://www.theatlantic.com/magazine/archive/1963/08/why-land-on-the-moon/361529/.

Our Horizon Goal: Mars (and Beyond?)

The Moon is an important stepping stone with geopolitical, scientific and the potential for commercial benefits. It is entirely possible that some of us will remain there for decades—for example, experimenting with in situ resource utilization (ISRU) technologies, conducting science, or seeking to extend our terrestrial economy to incorporate utilization and capitalization of resources found on or below the lunar surface. What we learn there will not only create opportunities in cislunar space but open new discoveries and knowledge that will help us as we look toward Mars. The Moon is not an end goal, but a beginning—a next step enabling the migration of technology, heavy industry, and humanity itself off the Earth and into the solar system at a scale that is no longer constrained by a single planet, our original home.

The Pathways report concluded that Mars is the "horizon goal" for the human exploration of space, the destination upon which the aspirations of all international space programs converge, and the farthest viable destination for human beings given foreseeable advances in technology. It is essential that this be an international effort, led by the United States in collaboration with others. The International Space Station has taught us that a multilateral enterprise such as Mars will bring forth intellectual capital, scientific abilities, research, engineering and interest in peaceful technology on the part of many nations. An international human Mars program, led by the United States, will build and expand on the foundation created by the ISS as well as lessons learned at the Moon.

At the same time, pushing further into deep space than humans have ever gone before offers the potential for technology breakthroughs—just as it did 50 years ago—unleashing American industry and investment in new and powerful ways. The fundamental barriers and challenges to planning and executing all large-scale enterprises—exploring space, mitigating climate change, controlling disease, or managing the rising global demand for clean water—are similar, whether on Earth or in space. Two of the most important needs in space—the availability and processing of water for life support and eventually propellant, and energy needs—obtaining, storing, managing, and transmitting energy or power—have direct relevance to many problems facing today's world. With thought and proper planning, technology opportunities and challenges on the Moon and particularly at Mars with the constraints imposed by its vast distance from Earth can find natural alignment with many of humanity's most pressing terrestrial problems.

In Closing. . .

Perhaps most importantly, our Nation needs the next generation of young scientists and engineers to advance our quality of life and remain globally competitive. The citizens of the United States also need a far better understanding of science and technology in order to exercise fully the rights and responsibilities of citizenship. The very future of our democracy depends on it. Nothing stimulates interest like truly great goals that require us to develop ourselves and advance the human condition as well as our technology in order to achieve them. Continuing the work begun with Apollo, returning to the Moon, and then reaching for the horizon of Mars, is just such a goal.

Senator CRUZ. Thank you, Dr. Dittmar. Mr. Hickam.

Statement of Homer H. Hickam, Jr., Author, Rocket Boys

Mr. HICKAM. Good afternoon, Chairman Cruz, Ranking Member Sinema, and Members of the Subcommittee, with a special shout- out to Senator Capito from my home state of West Virginia, who's probably thinking to herself, "Oh, my gosh, what's Homer going to say now?"

[Laughter.]

Mr. HICKAM. What Homer's going to say now is that I believe most firmly that our destiny lies at the Moon. We need to go back to the Moon. We need to put an anchor there and from that anchor develop the Moon commercially and scientifically.

One of the stories left out of the movie October Sky, which was based on Rocket Boys, was the fact that I met Senator John F. Kennedy when he was running for President in the West Virginia primary that he had to win, and since I was a rocket boy headed toward the National Science Fair, he asked me what we should do in space, and I answered Senator Kennedy. I looked around at all the coal miners in the crowd and I said, "Well, I think we should go to the Moon and just mine the blame thing," and all the coal miners, they just laughed and cheered, and he said, "Elect me President and maybe we will."

Most recently, one of my books called Back to the Moon, Vice President Pence told me that that's one of his favorite books and we talked at length and extensively about that. The next thing I knew, he was down in Huntsville, Alabama, at the Space and Rocket Center, at Space Camp, saying, "We're going back to the Moon by 2024."

So I'm kind of in the unique position of being able to take credit for both the Apollo——

[Laughter.]

Mr. HICKAM. — and the Artemis Space Program.

I do, I really do advocate going back to the Moon. There's a lot of work to do there. There are decades of work yet to be done.

Where was I when Apollo 11 landed? I was in the hospital ward at Ft. Lewis, Washington, with some fellow veterans of the 4th Infantry Division, and we watched the Apollo land. At that point, even though my mind had been occupied a great deal by Vietnam and all of that struggle, I realized I really wanted to work for NASA. I was going to make that happen.

It took me a while. 1981, when the shuttle started to fly, I got to work for NASA and I loved it. I woke up every morning and said to myself, "Oh, my goodness, this is great, I get to go work for NASA today." It's a tremendous agency and I absolutely loved it. Well, here we are and what are we going to do? Well, Mark Twain once said, "God looks after fools, drunks, and the United States of America," and that comment was never more apt than the way our space program has evolved. Modern American industry is now moving forward in space in astonishing ways and just SpaceX and Blue Origin are just a couple of those bright new companies that are coming up. They're able to come up with new designs and be more timely and create much less expensive ways to get in space. They're even beating out the heavily-subsidized space programs around the world, like China's. So we have every right to be proud of that.

Why are they able to do that? Because of a number of parallel forces. They include modern computer systems, the rise of the internet, new manufacturing techniques, and along other parallel lines, we see a demand for clean and abundant energy and awareness that Earth is in danger of being permanently polluted, and the gradual development of a new philosophy that seeks to put mankind in context with the universe.

As all engineers know, parallel lines never intersect until they do. These parallel lines of force I just mentioned intersect primarily at one place; our Moon, Luna, the small planet that circles us, our 8th continent, torn from the Earth a millennial ago, and so it is at this moment given to us by the benevolent hand of Creation that all of the elements needed to go back to the Moon and set up shop have come to us so as to utilize its mineral wealth and discover all that is there. And this includes, by the way, the possibility of evidence of life.

We've never looked at a single water molecule from space. We're going to find a lot of water ice on the Moon and in there might be evidence of life because the Earth and Luna have been sharing DNA for millions of years because of meteors and asteroids, sort of God's little space program. It didn't work out for the dinosaurs but it may work out for us.

In summary, I applaud the essentials of the Artemis Program to take us to the Moon, but I do say so with this proviso, that what we do on Luna must make sense to the American people both economically and philosophically, and should be designed in such a way that it may cost the people's money to place an anchor there but not so forever in order to sustain our presence.

The riches on the Moon, rare earth medals, thorium, titanium, helium3, and other minerals, should be gathered to boost our economy and thus put money in the pockets of all Americans. Our citizens should be assured that a base on the Moon will make this country stronger and safer, and I have to say I don't really much care who the next professional astronaut on the Moon is. What I care about who the next American plumber, electrician, construction worker, and blue collar worker on the Moon because when that happens, we know that we are finally a space-bearing nation.

Thank you.

[The prepared statement of Mr. Hickam follows:]

Prepared Statement of Homer H. Hickam, Jr., Author, Rocket Boys

Good afternoon, Chairman Cruz, Ranking Member Sinema, and members of the Subcommittee: It is a privilege to come before you today to speak on this vital topic during an exciting and dynamic new era of space flight that I hope will result in a permanent American anchor on the moon.

You may know me as the fellow who wrote the #1 New York Times bestseller Rocket Boys: A Memoir which told the story of some boys in the West Virginia coalfields of the 1950s who were inspired by Sputnik to build rockets. As you may also know, they made a movie titled October Sky based on that memoir which is still shown everywhere, especially in classrooms. That book and film has inspired many tens of thousands of young people to become engineers and/or work in the space business. For a West Virginia boy, son and grandson of coal miners, this is the most pleasing result possible for my work.

It should be noted that I have long been an advocate for a permanent American presence on the moon. This goes back to 1960 when I encountered then-Senator John F. Kennedy in McDowell County, West Virginia. At the time, I was 17 years old and heading for the National Science Fair with my rocket designs. He was running in the West Virginia primary for the Democratic nomination for President, a primary he had to win. After one of his speeches, he asked for questions and I asked him what he thought we should do in space. This was not a question I believe he expected in the coalfields and so he asked me what I thought we should do and I said we should go to the moon. Then, looking around at the coal miners in the crowd, I said we should just go up there and mine the blame thing. The miners cheered at this and Kennedy joined in, saying, "Well, elect me president and maybe we will!" He won the primary. It is why—very wryly to be sure—I sometimes take credit for the entire Apollo program although I know he was much more influenced by others such as Wernher von Braun.

It was my boyhood dream to get into the space business. However, this dream took awhile to happen. There are moments in our lives that beg the question: "Where were you when?" Such moments include the landing on the moon of Apollo 11. For me, I was in a hospital ward at Fort Lewis, Washington to watch the landing with other veterans after, in my case, a tour of duty with the 4th Infantry Division in the Vietnamese Central Highlands. President Kennedy, you see, had two important policies during his tenure, one of them was to see an American land on the moon and return safely, and the other was the defense of freedom around the world. Like a lot of young men of my era, it was the latter that occupied most of my life in the late 1960s and not the marvelous flights to the moon.

Eventually, I would come to work for NASA. This was in 1981 when the space shuttles began to fly. I was privileged to work at the Marshall Space Flight Center in Huntsville, Alabama, where I helped design the Spacelab, a precursor to the International Space Station, and train the astronauts who flew aboard the shuttle for Spacelab missions. I was also on the team that trained the astronauts underwater on the deployment of the Hubble Space Telescope and then later on the rescue and repair of that magnificent observatory. I trained the first Japanese astronauts and worked on the team assigned to fix the solid rocket motors that caused the Challenger tragedy. At the end of my career, I was on the team that negotiated with the Russians on how to train astronauts and cosmonauts to work aboard the International Space Station. I had a wonderful career and woke up every morning and said to myself, "Oh, boy! I get to go to work for NASA today!"

But I always felt like I had missed out on the great adventure that was going to the moon. I also had the sense that our country had foolishly and irrationally thrown away what could have been great and wonderful things for us if we had only stayed the course on what is essentially our eighth continent.

While I still worked for NASA, talking about the moon was pretty much off-limits but going against this policy, I stubbornly wrote a paper in 1993 on how we could eventually return to the moon and construct a base there. I have provided that paper to Senator Cruz's staff and it is available for your interest since much of it is still applicable, technically as well as philosophically.

After retirement in 1998, I took up writing full-time which resulted in Rocket Boys/October Sky and a bunch of other books including four set on the moon, one of which is titled Back to the Moon and, as it happens, one of Vice President Pence's favorite novels. Because of our mutual interest, I have talked with the Vice President at length about going to the moon and establishing a permanent presence there. I found the Vice President receptive and engaged and, as the board chair this year of the United States Space and Rocket Center (home of Space Camp) in Huntsville, Alabama, I was privileged to welcome him and the National Space Council in March, 2019 where he gave the speech that launched Artemis designed to take us back to the moon, this time to put down roots. In my own small way, I think I may have connected the timeline between the JFK and Vice President Pence decisions that the moon is where our destiny lies in space.

Incidentally, the Vice President's speech also directed something else, little noted or remarked upon but extremely important. He gave Marshall Space Flight Center in Huntsville the lead to develop nuclear rocket engines. If we ever want to send humans beyond the moon or want to send robotic missions very quickly far, far away, we will need a big boomer and not the relatively puny chemical systems of today. Boomers and other advanced propulsion systems are the future and I hope Marshall takes up the challenge. This would be with nuclear thermal propulsion, by the way, utilizing only low-enriched uranium so it is quite safe.

Mark Twain once said, "God looks after fools, drunks, and the United States of America," and that comment was never more apt than the way our space program has lately evolved.

Modern American industry is now moving forward in space in astonishing ways. Bright new companies—SpaceX and Blue Origin are only the best known—are outstripping not only NASA in innovative designs and much more timely and much less expensive ways to get into space but they are besting the heavily subsidized space agencies of the world. This is because of a number of parallel forces that include modern computer systems, the rise of the Internet, new manufacturing techniques and materials, and the imaginations stirred by popular science fiction films and books. Along other parallel lines, we see a demand for clean and abundant energy, an awareness that Earth is in danger of being permanently polluted, and the gradual development of a new philosophy that seeks to put humankind in context within the universe.

As all engineers know, parallel lines never intersect—until they do. Those parallel lines of force I just mentioned intersect primarily at one place, our moon—Luna, the small planet that circles us, our eighth continent torn from the Earth millennia ago, our friend in the sky without which life would not have evolved on Earth and now might sustain that life for centuries to come.

And so it is at this moment given to us by the benevolent hand of Creation that all the elements needed to go to the moon and set up shop have come to us so as to utilize its mineral wealth, and discover all that is there. This includes the possibility of the evidence of life in the water ice that we now know is so abundant there; for it is well known that the Earth and Luna have been sharing their DNA for billions of years through meteor and asteroid strikes which might be called God's space program. Our rendezvous with destiny on the moon is at hand.

Let me briefly speak of Mars. It is my belief that Mars, which is often mentioned by NASA officials and others as the ultimate destination for its astronauts and the moon only a stop on the way, should stay in the future for now. The truth is the technology to go there doesn't yet exist and for the blood and treasure it would take to put humans there, we could place thousands of robots on its surface and eventually artificial intelligence. Mars is interesting scientifically but it's the moon that should be our focus for the next several decades because, not only is there yet some great science to be done there, Luna has the potential of creating an economic powerhouse for our country and the world.

In that regard, I have to say I have little interest on who the next professional astronaut on the moon might be. My interest is who will be the first plumber, electrician, miner, or construction worker because when our blue collar workers arrive on the moon, we will truly and finally be a spacefaring nation and see all of our citizenry fully engaged in this magnificent adventure.

In summary, I applaud the essentials of the Artemis program to take us to the moon but I do so with this proviso, that what we do on Luna must make sense to the American people both economically and philosophically and should be designed in such a way that it may cost the people's money to place an anchor there but not to forever sustain our presence. The riches on the moon—rare earth metals, Thorium, Titanium, Helium-3, and other minerals—should be gathered to boost our economy and thus put money in the pockets of all Americans. Our citizens should also be assured that a base on the moon will make this country stronger and safer while also gathering knowledge for all the world so that humanity might rise above the immaturity evidenced by our endless conflicts, and become, as intended by our Creator, the responsible, mature adults of the Universe we were sent forth to be.

Thank you very much for your time. I look forward to your questions.

Senator CRUZ. Thank you, Mr. Hickam, and when you invoked the Twain quote about fools and drunks, I assume most people at home just assumed you were talking about the U.S. Congress.

[Laughter.]

Mr. HICKAM. No, sir.

Senator CRUZ. Mr. Stallmer.

Statement of Eric Stallmer, President, Commercial Spaceflight Federation

Mr. STALLMER. I have a tough act to follow. Homer was there to inspire the Apollo Program. I wasn't even born for Apollo 11. So you raised the bar quite a bit.

Chairman Cruz, Ranking Member Sinema, and Distinguished Members of this Committee, thank you for inviting the Commercial Spaceflight Federation today to discuss how rapid advances by U.S. commercial space industry can help NASA truly honor the Apollo legacy by leading humanity into the Solar System further and faster.

In the past two years, NASA has crystallized an ambitious agenda to commercialize low-Earth orbit, establish a long-term presence on the surface of the Moon, and send humans to Mars. These goals are all linked together and collectively they will enable an expansion of our civilization in the Solar System.

It is the breadth of this vision of space as a frontier for all that makes the recent emergence of the strong U.S. commercial spaceflight industry so uniquely valuable. It didn't just happen overnight.

For two decades, NASA has fostered the development and increasing success of this industry, sharing technologies and expertise, co-investing in private innovation and using its purchasing power to serve as an initial customer.

Those steps have enabled private companies to develop, own, and operate their own human spaceflight hardware to serve the public needs as well as the private sector markets.

These efforts were established in NASA's mandate to seek and encourage, to the maximum extent possible, the fullest commercial use of space, as well as the bipartisan legislation and administrative policies under both Democratic and Republican presidents.

Over just the past two years, Space Policy Directives 1, 2, and 3 have strengthened this partnership between government and industry and helped remove barriers to industry growth, and under your leadership, this Committee has worked to facilitate industry's development via the Space Frontier Act. We thank you.

At the start of the Space Age, the United States established its leadership in space with government-funded and led exploration projects. When President Kennedy proposed sending an American to the Moon, there were no alternatives to the all-government strategy.

But today, five decades after Apollo, the United States is enjoying a renaissance in space with commercial space enterprises playing a leading role. The details are in my written testimony but in summary, the U.S. has leapt forward in 2018 in every commercial space sector with more launches, more spacecraft pursuing both old and new space applications benefiting more and more stakeholders here on Earth.

So today, as we seek to further develop low-Earth orbit, place an enduring American presence on the Moon and sending not just two but many brave pioneers to Mars, policymakers like yourself have newer, more affordable, and sustainable options than just repeating Apollo.

The ingredients of a successful strategy have proven themselves over and over in recent years. Future LEO infrastructure, every operational element of returning to the Moon, and most of any of any affordable Mars architecture should be purchased commercially or developed via COTS-like partnership.

For example, allowing the private sector to develop its own solution for the Lunar Gateway Power Propulsion Element, NASA was able to generate a fixed price bid that was more than $200 million less than the closest competitor.

While science is appropriately a government-led activity, the basic engineering and infrastructure that supports it can be provided commercially with only the leading edge tools and instruments required by government stewardship.

NASA should specify clear high-level outcome-based requirements and allow entrepreneurs to innovate and create affordable and basic capabilities to meet essentially all the operational needs, and NASA must pay for results, not effort, on all developmental programs but the most esoteric technical challenges.

Whenever possible, NASA should award multiple competitively chosen funded Space Act agreements to commercial partners willing to put up private capital at their own risk. That leverage, plus ongoing competition, will replace any need for the costly micro-management and bureaucracy of typical FAR-based contracts.

Competition also allows for greater diversity in technical approaches and much lower strategic program risk. None of this will be easy. Commercializing low-Earth orbit will be hard. Human travel to the Moon is hard and staying there will be even harder. Mars will be harder still but with even greater rewards as we explore a different planet and its moons.

But American industry is ready to help NASA chart an affordable and sustainable path into this challenging future.

This month, it's natural to venerate the past, but we should also be proud of the great new things we are achieving today and what we can do together tomorrow if we build a true partnership between government, including Congress, and the American people on their enterprise.

Chairman Cruz, Ranking Member Sinema, thank you for your invitation and attention. I look forward to your questions.

[The prepared statement of Mr. Stallmer follows:]

Prepared Statement of Eric Stallmer, President, Commercial Spaceflight Federation

Chairman Cruz, Ranking Member Sinema, and distinguished members of the Committee—thank you for inviting the Commercial Spaceflight Federation (CSF) to discuss our members' views on the state of the U.S. commercial space industry. We also appreciate the opportunity to honor the 50th anniversary of Apollo 11 by looking at NASA's exploration plans and examining how our past experiences and lessons learned can guide space exploration efforts going forward. In addition to NASA's world-leading space exploration capabilities, the United States now has a vibrant, highly capable commercial space sector that can accelerate and complement NASA's internal development efforts. As we look to the future, true public-private partnerships between this domestic industry and the Government represent the most effective path to rapidly, safely, and cost-effectively return to the Moon and venture on to Mars.

CSF is the leading national trade association for the commercial space industry, with more than 85 member companies and organizations across the United States. Founded in 2006, CSF is focused on laying the foundation for a sustainable space economy and democratizing access to space for scientists, students, civilians, and businesses. CSF members are responsible for the creation of thousands of high-tech U.S. jobs driven by billions of dollars in investment. Through the promotion of technology innovation, CSF members are guiding the expansion of Earth's economic sphere, bolstering U.S. leadership in aerospace, and inspiring America's next generation of engineers, scientists, and explorers.

NASA has embarked on an ambitious effort to commercialize low-Earth orbit (LEO), to establish a long-term presence on the surface of the Moon, and to send astronauts to Mars. These bold commitments should be commended. Over the last two decades, NASA has fostered a nascent domestic spaceflight industry into becoming a highly diverse and capable portfolio of companies. NASA has invested in private development, used its purchasing power to serve as an anchor customer, and enabled private companies to develop, own, and operate their own human spaceflight hardware to serve both public and private needs. Because of the agency's foresight and meticulous cultivation of this industry, American companies support critical space exploration and national security needs today, in addition to the commercial marketplace.

Policymakers have recognized the benefits of these kinds of partnerships since the earliest days of the space program. The National Aeronautics and Space Act of 1958 outlines one of NASA's core missions is areas as: "[t]o seek and encourage, to the maximum extent possible, the fullest commercial use of space." National Space Policies from both Democratic and Republican administrations have stressed the importance of the commercial space sector. And Space Policy Directives 1, 2, and 3, each issued over the past two years, take further steps to strengthen the partnership between Government and industry and to remove barriers to industry growth. And, under your leadership, this Committee has taken important steps to facilitate commercial space industry development, most recently with its efforts on the Space Frontier Act.

Today, I will outline CSF's perspectives as to how we all can collectively advance our Nation's space goals through innovative, strategic partnerships with American industry.

I. America's Vibrant, Highly Capable Commercial Space Sector

The United States established its leadership and dominance in space with government-funded and government-controlled space exploration.[14] When President Eisenhower founded NASA and President Kennedy outlined a goal to send Americans to the Moon, there was no other choice; NASA literally had to invent whole new fields of technology, not just new hardware. The agency's accomplishments are a marvel for the ages, progressing from sending an American into space for the first time in 1961 to landing a crew on the Moon just eight years later.

Thanks in large part to NASA's leadership, pathfinding, and partnerships with the private sector in the decades since, a broad and dynamic space industry has emerged. Since 2000, investors have supported 375 private space companies with nearly $19 billion of private capital.[15] As NASA continues to drive the frontier outward with groundbreaking research in space, the commercial sector is making space affordable and accessible to everyone.

Today, the United States is enjoying a renaissance in space, with commercial space enterprises playing a leading role. To update the Committee on the commercial space industry's recent major milestones:

- Last year, U.S. commercial space companies achieved an unprecedented 32 licensed orbital and suborbital launches as well as 14 licensed reentries. SpaceX conducted the majority of those licensed activities, with 21 launches and 12 first stage landings. American commercial providers of medium-to-heavy lift launch services now represent a supermajority of global commercial launches each year.
- Over the past several years, there has been a surge of progress from dedicated small orbital class launch vehicles. In 2018, Rocket Lab conducted the first successful launch of its Electron rocket. Rocket Lab has already launched three more times in 2019, orbiting 35 satellites—including two for U.S. Special Operations Command. Relativity Space is building an autonomous rocket 3D printing factory in Mississippi, expanding capabilities at NASA's Stennis Space Center. Vector Launch and Vox Space (a Virgin Orbit

[14] Space, The Final Economic Frontier, Journal of Economic Perspectives, Matthew Weinzierl, Spring 2018. Available at: https://pubs.aeaweb.org/doi/pdf/10.1257/jep.32.2.173.

[15] Proprietary Data, Space Angels, September 30, 2018.

subsidiary) have been selected to compete for DARPA's Launch Challenge. And, Virgin Orbit has completed several captive carry tests of its LauncherOne vehicle attached to a 747 aircraft in preparation of flights to space in the near future.

- A growing number of companies are restoring and expanding America's human spaceflight capabilities. This year SpaceX—in close partnership with NASA— will launch American astronauts to space in an all-American system, ending the country's drought on orbital human spaceflight capability left by the retirement of the Space Shuttle in 2011. Already, SpaceX and NASA conducted a successful flight qualification mission of the Crew Dragon spacecraft in March. Virgin Galactic successfully launched three spaceflight participants on its spacecraft— SpaceShipTwo—into space for the first time, reaching an apogee of 51.4 miles. Blue Origin has conducted a series of uncrewed suborbital test flights on its New Shepard vehicle and plans to conduct a test flight with crew soon. Both companies plan to fly spaceflight participants to space for revenue by the end of the year.
- American companies continue to make significant progress commercializing the International Space Station (ISS) and LEO.
    - Sierra Nevada Corporation's (SNC) Dream Chaser spacecraft—in an uncrewed cargo configuration—passed a key milestone in its development to be the third commercial cargo vehicle for the International Space Station.
    - Nanoracks has supported more than 750 payloads at the ISS to-date and has deployed 243 satellites through a commercial dispenser, in partnership with NASA. The company is also building the first-ever commercial airlock, designed to be integrated with the ISS in the coming years.
    - The ISS National Lab has facilitated more than $150 million in external, non- NASA funding to support the full ISS National Lab portfolio—a 50 percent increase in FY18.
    - Axiom and Bigelow are developing commercial space habitats, and each has made major technical progress over the past year.
    - Made In Space, TechShot, and Space Tango continued to demonstrate additive manufacturing and other interesting commercial applications in microgravity.
- Planet, Blacksky, and Maxar Technologies deployed dozens of new commercial remote sensing satellites to orbit.

- Southwest Research Institute (SwRI), University of Colorado, Boulder, and Maxar Technologies were selected by NASA to build three new lunar science and technology payloads to fly on future flights through NASA's Commercial Lunar Payload Services (CLPS) project. NASA selected three commercial Moon landing service providers that will deliver science and technology payloads under CLPS as part of the Artemis program.
- The first licensed flights to space of two American suborbital reusable launch vehicles: Blue Origin's New Shepard and Virgin Galactic's SpaceShipTwo. I emphasize the word licensed, because a license allows the company to earn revenue from the flight, unlike an experimental permit.
- World View performed its longest flight to date of its stratospheric balloon, demonstrating its ability to carry out missions traditionally reserved for satellites.
- And, in May, NASA has entered into partnerships with 11 companies—including CSF members Blue Origin, SpaceX, Sierra Nevada, and SSL (a subsidiary of Maxar Technologies)—to conduct advance development on human lunar lander concepts.

These recent achievements are just a few of many by the commercial industry, and they set the stage for even greater accomplishments the rest of this year and beyond for a broad set of stakeholders.

II. Apollo 50th: Partnerships with Commercial Industry Are Fundamental to Achieve a Sustainable Return to the Moon

50 years after Americans first stepped on the surface of the Moon, President Trump, NASA and Congress have established a national commitment to return Americans to the Moon—not just to plant footprints and flags, but to establish long-term habitation and sustainable activity on the lunar surface. This will provide the spaceflight community a valuable proving ground for NASA's goal of sending astronauts to Mars.

This ambitious objective should be applauded and, if executed appropriately, will serve to reinforce American leadership in space as international competitors like China, Russia, and India focus their own exploration efforts on the Moon. In January 2019, China achieved a major milestone in its lunar space exploration program, landing the Chinese space agency's Chang'e-4 spacecraft on the far side of the Moon for the first time in history.[16] This achievement builds on the China's 2013 success of landing its first rover on the Moon, joining the United States and the Soviet Union as the only nations to have carried out a soft landing on the Moon. In September this year, India hopes to become the fourth country to soft land on the Moon.[17] The United States should ensure it remains the leader in space exploration—and private industry is here to help.

Long-term, sustainable exploration on the lunar surface, and ultimately Mars, requires an integrated effort that includes the development of capable landers, the operation of robust deep space habitats, and routine transportation of astronauts and large cargo. Public-private partnerships with commercial companies are fundamental to developing these capabilities. NASA's Fiscal Year 2020 budget request and budget amendment prudently highlight partnerships with commercial providers as a key tenet of this strategy.

Undeniably, the systems that brought Americans to the Moon during the Apollo program were and remain a marvel. But, as we consider our options for the future, we should look not only to technologies of the past, but to the new advancements of today. CSF companies are proud to be playing a role in this new era.

III. Flexible COTS-like Development Agreements and Firm, Fixed-priced Services Contracts

These capabilities are already helping to support NASA's exploration goals, and they will continue to support NASA as it works to return to the Moon. As this Committee looks to how best ensure the country's ongoing leadership in space, it must carefully review development and acquisition efforts to ensure responsible use of finite taxpayer dollars and to encourage, rather than hamper, rapid innovation.

True commercial partnerships for development and operation of some elements of the exploration architecture represent the most rapid and cost-effective path to return to the Moon. In these partnerships, NASA outlines high-level mission objectives and safety requirements, but does not dictate

---

[16] The New York Times, "New Chapter in Space Exploration as China Reaches Far Side of the Moon," January 2019. Available at: https://www.nytimes.com/2019/01/02/world/asia/china-change-4-moon.html.

[17] CNN, "India hopes to become fourth country on the moon in September," June 2019. Available at: https://www.cnn.com/2019/06/13/india/india-moon-mission-intl/index.html.

system designs. Companies are required to compete for awards and to self-invest; and they are paid on a fixed-price basis only upon achieving milestones. Further, these industry-led partnerships allow NASA to be one customer of many, stimulating a vibrant, commercial lunar economy. Already, due in part to the stability that NASA brings to the market as a customer, numerous private companies are developing lunar systems and signing commercial contracts with customers around the world.

NASA should consider the Commercial Cargo program and its development effort—Commercial Orbital Transportation Services (COTS)—as the ideal model as it looks to structure the lunar lander and habitat elements of its exploration architecture. By even the most conservative independent evaluation, the COTS Commercial Cargo public-private partnership saved the agency hundreds of millions of dollars and allowed NASA to redirect those savings towards funding its other priorities, including earth observation and deep space exploration.

Numerous independent reviews of the program have repeatedly praised this partnership for its significant savings for the taxpayer. In August 2011, NASA, using the NASA-Air Force Cost Model (NAFCOM), determined that had the agency saved between $1 billion and almost $4 billion by using the COTS model as compared to a traditional procurement approach.

A 2014 NASA report further praised the program's use of innovative, flexible Space Act Agreement (SAA) development arrangements: "[b]ecause these were partnerships, not traditional contracts, NASA leveraged its $800M COTS program budget [for both providers combined] with partner funds. This resulted in two new U.S. medium-class launch vehicles and two automated cargo spacecraft and demonstrated the efficiency of such partnerships."[18] A 2017 NASA Cost Analysis review was more direct: "the COTS development and later the operational Commercial Re-supply Services (CRS) are significant advances in affordability by any measure."[19] Simply put, this approach works.

As NASA looks to define a development and acquisition approach for its lunar lander and habitat systems, it should adopt a COTS-like structure that would:

---

[18] NASA, ''Commercial Orbital Transportation Services: A New Era in Spaceflight,'' February 2014. Available at: https://www.nasa.gov/sites/default/files/files/SP-2014-617.pdf.

[19] Zapata, Edgar. An Assessment of Cost Improvements in the NASA COTS/CRS Program and Implications for Future NASA Missions. American Institute of Aeronautics and Astronautics, 23 Oct. 2017, https://ntrs.nasa.gov/archive/nasa/casi.ntrs.nasa.gov/20170008895.pdf, pp. 1.

- Leverage the commercial industry's ability to innovate quickly, improving safety and reliability. The sector's high cadence of development and test-like-you-fly approach provides for far greater system maturity than relying purely on simulations and ground tests. This approach was what NASA followed during the Apollo era, and it should return to its roots.
- Establish clear, high-level, milestone-based requirements that enable creative, innovative, and cost-effective solutions and avoid overly-specified and everchanging Government requirements. This structure forces the Government customer to get the requirements right and clearly communicate priorities at program start.
- Use firm, fixed-priced, pay-for-performance, milestone-based agreements that drive toward a successful conclusion and focus on an outcome-oriented commercial service. This commercial structure incentivizes companies to provide the deliverable at the time, place, and price negotiated with the Government, and discourages continuous Government requirement changes that add costs and delay schedules.
- Maximize competition throughout the entirety of the program. Competition is critical to accelerating progress, driving value and performance, and improving the quality of service to the customer. Price competition obviates the need to levy expensive, anti-competitive, non-value added requirements for certified cost or pricing data.
- Require a significant private capital contribution to the overall program. Commercial partners should share costs and provide a significant percentage of the overall investment, resulting in lower costs to the Government and enabling it to stretch its budget further.
- Tolerate programmatic risk and allow easy termination for failure to meet early requirements. The Government needs the flexibility to terminate contracts and cut bad actors when programs go far over budget and behind schedule.
- Encourage new, non-traditional companies to work with NASA. Traditional FAR-based contract requirements are complex and costly, which often deters small, less-experienced companies from working with the Government. As a result, the Government is often not at the cutting edge of new commercial technology offerings. The use of

COTS-like contracts can help enable such companies to do business with the Government.

- Facilitate the development of new markets and leverage market-driven pricing to support Government requirements and missions.

Not only must NASA plan prudently to save money, it likely must also anticipate and plan for funding levels below its requests to Congress, due to budget issues entirely unrelated to the agency. Indeed, NASA is already anticipating and planning for such a scenario.

When faced with budget shortfalls, NASA often attempts to make up for the shortfalls by: 1) drawing funding one part of the agency to pay for another part of the agency; and 2) deferring, de-scoping, or discontinuing lower priority programs and activities within the agency. Both options are demonstrably bad choices and lead to even worse results for the agency—undermining support for the Moon initiative, destabilizing other programs and missions, and leading to increased costs and schedule delays across the agency.[20] Instead, the Committee should encourage NASA to follow a third way that avoids these pitfalls: public-private partnerships.

True commercial partnerships for development and operation of some elements of the exploration architecture represent the most rapid and cost-effective path to return to the Moon. Pay-for-performance creates the proper incentives on both sides of the Government/company relationship. Here, the GAO has reported: "[f]irm-fixed-price contracts place the onus on the contractor to provide the deliverable at the time, place, and price negotiated by the contractor and the government. In addition, firm-fixed-price contracts place the maximum risk on the contractor as well as full responsibility for all costs and any resulting profit or loss."[21]

To succeed, NASA must employ fast, flexible, lean contracting agreements like SAAs to incentivize rapid and affordable development of U.S. transportation and habitat systems to safely land humans on the Moon by 2024.

---

[20] NASA Office of Inspector General, "NASA Cost and Schedule Overruns: Acquisitions and Program Management Challenges," June 2018. Available at: https://oig.nasa.gov/docs/ CT-18- 002.pdf.

[21] Government Accountability Office, "NASA: Acquisition Approach for Commercial Crew Transportation Includes Good Practices, but Faces Significant Challenges," December 2011, (GAO–12–282). Available at: http://www.gao.gov/assets/590/587021.pdf.

IV. Lessons from Apollo: Constant Innovation and an Eagerness to Test New Concepts

By any accounting, the Apollo program represents humankind's greatest and most inspirational technological achievement. It was an enormous undertaking, costing about $177 billion in 2019 dollars. Only the building of the Panama Canal rivaled the Apollo program's size as the largest non-military technological endeavor ever undertaken by the United States and only the Manhattan Project to build the atomic bomb in World War II being comparable in a wartime setting.[22] As we honor the 50th anniversary of Apollo 11, CSF commends the Committee for examining how our experiences and lessons learned can guide space exploration efforts going forward.

Several important lessons from Apollo need to be remembered and should guide space exploration efforts going forward. Iterative, evolutionary, and risk-tolerant development was the cornerstone of NASA's Apollo-era progress:

- From 1958 to 1963, the Mercury program conducted twenty uncrewed development test flights, and six successful flights with astronauts. Mercury's early uncrewed test flights helped NASA find and fix problems. For example, of the twenty uncrewed test flights, half of the flights resulted in failure or partial success. But, the agency learned how to put astronauts in orbit around Earth. It learned how people could live and work in space, and it learned how to operate a spacecraft in orbit.
- Following Mercury, during a 20-month span of the Gemini program from March 1965 to November 1966, NASA flew ten Gemini crews to Earth orbit, each testing new capabilities in preparation for the landing on the Moon.
- And, of course, Apollo followed these programs. From 1961 to 1966, NASA launched increasingly capable iterations of the Saturn rocket 14 times to learn valuable insights into the complexities of building and launching a large cryogenic system so that, before the decade was out, they had the capability to land people on the Moon.

The Apollo-era attitude that enabled NASA to land on the Moon 50 years ago was based on constant innovation and an eagerness to test new concepts.

---

[22] https://history.nasa.gov/Apollomon/Apollo.html.

These are the same principles that underpin the success of the commercial spaceflight industry today. For example:

- Small-, Cube-, and Nano-Satellites are increasingly effective and efficient platforms for remote sensing and communication applications, and commercial providers developing and operating these systems are revolutionizing the satellite industry.
  - From 2013 to 2018, Planet, an Earth imaging company, launched nearly 300 satellites into orbit, 150 which remain active. Combined, they constitute the largest constellation ever put into orbit.
  - Planet is not alone, companies like Advanced Space, Amazon, Astranis, BlackSky, Maxar Technologies, and SpaceX are each developing their own systems, with SpaceX launching the first 60 satellites of its Starlink broadband constellation this past May.
  - Thanks to the advances by the commercial space industry, NASA is now utilizing these platforms to augment and support the scientific investigations carried out by the agency's larger flagship missions.
- Suborbital reusable launch vehicles are providing low-cost and frequent access to suborbital space for humans and research payloads that are contributing to NASA's science, exploration, and technology development missions. During a six-month span from December 2018 to May 2019, Blue Origin and Virgin Galactic flew to space and back four times, carrying 54 research and technology payloads, 24 of which were NASA-sponsored payloads testing new capabilities for the agency.
- The commercial industry has established and self-funded the world's first marketplace of diverse reusable rockets, both suborbital and orbital, which successfully launched for commercial and government customers.
- Commercial landers: successfully hotfired large lander engines, which have been under development for several years, and exclusively developed with private capital;
- Commercial heavy lift: successfully launching the world's most powerful operational rocket by a factor of two, exclusively developed with private capital; and successful test-firings of large methane rocket engines and development of heavy lift launch vehicles;

The need for flexibility serves as another important lesson for Apollo. For example, NASA was the was the first Federal agency to be granted Other Transaction Authority (OTA) in the 1958 Space Act, in order to provide NASA with the full flexibility to beat the Soviets in the space race to the Moon.[23] Now, as then, NASA must adapt when new technologies and architectural approaches are introduced.

V. Conclusion

With the technological advancements and increased knowledge achieved through decades of work by NASA in deep space, including Mars, the United States is now well-positioned to build upon and surpass our past achievements in space. With NASA resources and expertise, coupled with American ingenuity, the principles of free enterprise, and the benefits of competition, the United States can do more in space than has ever been accomplished previously. We just need to appropriately recognize and leverage our advantages.

As the Nation commits to returning to the Moon by 2024, it is buoyed by a vibrant commercial space industry powered by agile and innovative development processes, flexibility and some level of risk tolerance, private capital co-investment, and more intensive innovation. In its return to the Moon, NASA does not need to go it alone, nor should it. The most efficient and realistic way of returning to the Moon is a hybrid approach between Government and commercial partners.

Going to the Moon is hard. Staying there is harder, and moving beyond to Mars is harder still. But American industry stands ready. The Commercial Spaceflight Federation supports NASA's Fiscal Year 2020 budget request and budget amendment to more fully utilize public-private partnership programs and commercial service buys to accelerate cost-effective deep space exploration objectives, including sending landers and astronauts to the lunar surface by 2024.

---

[23] "This was 1958 and NASA was a big deal," said Ralph Nash, a government procurement law expert and founder of the Government Contracts Program at George Washington Univer- sity's National Law Center, where he is a former dean of graduate studies and professor emer- itus. "We were in a space race with the Russians and President [John] Kennedy said we would get to the moon in this decade,'' he explained. 'This was there to give them full flexibility." See here: *https://www.nextgov.com/it-modernization/2018/03/otas-scary-new-contracting-model- isnt-scary-or-new/146964/*

It is now time for the United States to build on Apollo's important legacy. To do so, CSF recommends the following:

1) Use flexible development agreements like SAAs for development activities and firm, fixed-price contracts for services;
2) Focus procurement approaches and requirements on an outcome-oriented integrated commercial service rather than a government owned or operated systems;
3) Competitively awarded, firm, fixed-price contract with payment for meaningful deliverables and milestones, not just for effort;
4) Maintain competition throughout—two or more companies should proceed through the flight demonstration phase for each program element and into fol- low-on service phase;
5) Eliminate Cost Accounting Standards (CAS) when there is competition and fixed price contracting—CAS degrades speed and adds costs without improving contract performance. Fixed-price milestone contracts place risk on the contractor for costs and schedule, obviating the need for cost reporting elements. Cost Accounting Standards also serves as a barrier to entry for non-traditional firms, artificially limiting the competitive pool; and
6) Mirror commercial terms and conditions to the maximum extent. Eliminate all other FAR-derived provisions that are not essential to incentivizing the core outcome.

These are exciting times in spaceflight. We should all be proud of what the American space enterprise—both the Government and the private sector—is achieving. The challenges we face to achieve our goals today are not small, but we have the ability and opportunity to address them in a thoughtful and impactful manner given Congress' and the Administration's support.

Indeed, NASA's budget amendment states that to "achieve our goals, we will not go forward alone," and "strong commercial partnerships will accelerate our human exploration plans." We are ready to take that step with NASA, and we look forward to continuing to work with this Committee to the Nation sustainably returns to the Moon and ventures to Mars. Chairman Cruz, Ranking Member Sinema, I appreciate your invitation to testify before the Committee today. Thank you for your attention, and I look forward to your questions.

Senator CRUZ. Thank you, Mr. Stallmer. Thank you to each of the witnesses for your testimony today.

As we sit here today reflecting back on the last 50 years and the journey that America has traveled and that space exploration has traveled since that giant leap for mankind, I think it is also appropriate for us not just to look backward but to look forwards and to ask what do the next 50 years hold?

In the year 2069, perhaps our kids or grandkids will be participating in another hearing marking the 100-year anniversary of man landing on the Moon, and so my first question I want to ask the panel is in the next 50 years, what should we hope to accom plish in space? What should our objectives be? What should we be looking forward to?

Mr. STALLMER. I think what is paramount is that over the last 50 years, we could have and should have done a lot more. As young people that looked at the Apollo era 50 years ago and said, gosh, we landed a man on the Moon, in 1985, what are we going to do, and 20 years later what could we be doing, and what have we done? We haven't really returned back to the Moon as humans.

I think we really need to accelerate and not think as much the next 50 years but maybe the next 20 years what we can do and I think we have all the ingredients to do it properly.

I think this reinvigorated look to the Moon, to the exploration, to setting up a base on the Moon, what Gateway will provide, and the outlook and future vision for what we can do on Mars. I don't think we should focus so far in 50 years, but I think in the next 25 years, when a lot of these young people in this room behind me are coming of age and these engineers that they will have that opportunity to live and work in space and develop the new businesses that are going to help promote—I mean, the global space market right now is a $360 billion industry.

I agree with you that it's going to be a trillion dollar industry, probably within a decade, and we have to look at what are those commercial opportunities that are going to be available in space and that's what excites me.

Mr. KRANZ. I think from my standpoint, we had an excellent plan that was written by Tom Paine. It was probably the most widely distributed plan with participation from industry, commerce, academia, right on down the line. It was called Pioneering the Space Frontier.

I think the key thing that we have to do is establish a plan and stick with it. Since then, there have been at least three other plans. It was interesting in that Pioneering the Space Frontier, Gateway was discussed but never showed up in any of the subsequent plans. So the question is, is that there's good work out there. What we have to do, I think really this Administration should do to fit into the Artemis plans here, is truly take a look at those plans that have been written and see what parts still are viable, what parts fit together, because you're talking about a direction well beyond the Moon. You're talking about a direction for the future and this plan was really—the Pioneering the Space Frontier was a 50-year plan, I believe. I think it's worthwhile to address.

Senator CRUZ. Mr. Kranz, in your testimony, you stated that you don't see the same national unity that we saw in the 1960s that led to the successful Apollo Program.

Mr. KRANZ. Yes.

Senator CRUZ. Can you elaborate on that——

Mr. KRANZ. Well, this—

Senator CRUZ. — and what we need to do to achieve that unity once again?

Mr. KRANZ. I could talk for hours on this. I've finished writing a book. I don't know if it'll get published because I've addressed my perspective of the agency and our work in space since the time I joined back in 1963 through actually the first four decades. It was very influential from my standpoint.

I was trying to figure out why—I was director of Mission Operations for a while and why we got such confusing change of plans, directions, right on down the line. Aaron Cohen in the oral history described the period after Challenger as the time worse than chaos. He was one of the center directors. He'd been up at headquarters. That was his part of his oral history.

Dick Truly did the same thing. He talked about rough seas, steadying the course. So it's really a question that I go back into the leadership focus that we need within the agency but it starts at the very top and says this is what we're going to do.

What I want you to do is give me a set of plans to accomplish that objective. I'll work with you to pick the people but we need to find leadership. We ought to go to the—you know, when I was growing up in space, we had very competitive industries. We had incredible aircraft and aerospace industry in there. We were able to get people from McDonnell Aircraft, John Yardley to come down and show us how to start a program and get it started. He was teaching us the business of program management.

Dale Myers came in, the same thing. I hated the Falcon telcons, but they were places where we learned the business from the professionals of the business. So I think the key thing is the future is there. A lot of the plans are there. We have to establish which plan we're going to subscribe to and get on with it and then find the leaders to implement it.

Senator CRUZ. So final question. In your judgment, what are the benefits to Americans of returning to the Moon and establishing a sustainable habitat on the Moon? Mr. Hickam, you addressed this already.

Mr. HICKAM. Yes, you know, I was raised in a little town of Coalwood, West Virginia, and in places like Coalwood, bituminous coal was brought up in the early 20th Century that basically fueled the American economy and really the civilization of this country. So what would I like to see in 50 years? I'd like to see Coalwood on the Moon. I'd like to see families raised on the Moon. I'd like to see blue collar workers, miners, getting money in their pockets and essentially creating a space-based economy because I agree with you; the first trillionaire is going to be from space. The place to go to make that happen is the Moon, and how can we make that happen?

When Mr. Carter who built Coalwood first went in there, he went on the back of a mule but that mule went—first, he got off the railroad, took the mule and him on the railroad to a certain station, then got off of that on a road that was built that took him into Coalwood and there he sunk the shaft and brought up about a million tons of coal.

Then it's the government's responsibility, in my opinion, to do very much the same on the Moon. We need to put an anchor on the Moon, a place where people can go, companies can go, countries can go, and from there they can branch out and develop Coalwoods on the Moon and basically cause a real space economy to develop.

Senator CRUZ. Well, and I'm looking forward to seeing the space suit that we can fit a mule into.

[Laughter.]

Mr. HICKAM. Me, too.

Senator CRUZ. Senator Sinema.

Senator SINEMA. Thank you, Mr. Chairman. Thank you again for your testimony.

My first question is for Dr. Darden and Mr. Kranz. Both of you worked at NASA leading up to and during the Apollo missions and can offer a unique perspective on how NASA achieved its ambitious goal in the 1960s.

What lessons do you believe NASA's current leadership and workforce can learn from your experiences at NASA during the Apollo missions as we prepare to return to the Moon?

Dr. DARDEN. Well, when I went to NASA, they were two years away from walking on the Moon, I went in 1967, but I think his comments about the leadership, the commitment, and the background that they had in working with the other programs, Project Mercury and Gemini, I think all of those were factors of how these persons pulled together and were committed to making that a successful program, successful project.

Mr. KRANZ. From my point, I think that there was the mutual sharing of knowledge. When I first got to NASA, I told you I didn't know anything about rocket spacecraft, Mercury spacecraft, but I could walk around Hangar S where the spacecraft. I could go down to the Block House, talk to the booster people. I could go over to the Marine Safety people.

So it was a question. There was so much to do and so much to learn that literally everybody spent the entire day telling everybody what they learned and vice versa.

So I think we have to re-establish this passion, the energy, the imagination in our organizations, and I think this starts right at the top in the leadership. I think somewhere along the line our leadership values have sort of been subtly changed. It might be generational to a great extent, but I think the generation I came from, I call it the Apollo generation, the sons and daughters of the Greatest Generation, I think they had a, what I'd say, growing up, born in the Depression.

I think we have to sort of like—we constitute what our Nation is and what it stands for and what we expect from our people. What are the expectations we have for all employees at NASA, Federal Government, right on down the line?

And Mission Control, right after the Apollo 1 fire, which was disastrous, many of us lived through that 18 seconds, but we sat down after we finished and wrote up the searing impressions we had in such a fashion that we would never go through this again and we call it The Foundations. In fact, it's on the Internet and I get requests for this every day but I'm surprised going to Fortune 500 companies and see it on the wall.

I see the Kranz Dictum where I was angry, passionate, accept responsibility for the Apollo 1 fire. The crew could have called it off. I could have called it off. The touch conductors could have called it, but nobody said stop, it's not right.

So it's a question we have to redefine expectations that we have for ourselves, our organizations, our people, and our leadership, and I think it's there. What we've got to do is put it in words. We need to establish some shared values that we're all working to.

Senator SINEMA. Thank you. GAO's 2018 Assessment of Major Projects at NASA pointed out that more than half of NASA's workforce is over the age of 50 and 21 percent of NASA's workforce is already eligible for retirement.

Workforce challenges exist beyond NASA. Since the Shuttle Program ended in 2011, many of the commercial manufacturers and suppliers who had for decades supported NASA missions were left to close up shop and many of the skilled employees who worked to support NASA programs went and found work elsewhere.

So my next question is for Dr. Dittmar and Mr. Stallmer. As the United States prepares again to send Americans into space and return to the Moon, do you believe the current workforce at NASA and its commercial partners is sufficient to support a long-term series of American space leadership?

Dr. DITTMAR. So in my written testimony, I did address this to some extent, but let me just sort of recap briefly.

There have been several studies recently that have looked at when people first become interested in going into aerospace and about 71 percent of them become interested while they're in grade school.

The pressure that you're describing with regard to people continuing to work, stay in the workforce longer does make it difficult for younger people to enter. At the same time, however, there are terrestrially based activities, IT, computing, advanced artificial intelligence, a number of those, and those high-tech firms, they actually have been attracting for some time some of the best and brightest in the country.

So the situation that we're facing is we have an aging work force. We do have positions opening up but a lot of those positions are being competed for by other industries that are very attractive and so as a country, I think it's very important that we focus on making sure that we continue to provide educational opportunities, that that education is top-notch.

The United States remains in the third, the upper third by almost every measure I've seen in the last 10 years, we haven't moved very much, the upper third of other nations with regard to STEM education. Upper third is OK. It's better than the lower third, but it's not where we should be.

So I think a national focus on making sure that our folks are actually prepared to go into this work, continuing development of jobs, both through national programs and through commercial development, I think is very important, so we open up opportunities for people to be able to come into that.

The same is true for science. We underfund science in this country to an extraordinary extent. That's been noted for the last 20 years, since the Augustine Rising Above the Gathering Storm. Actually it's a little bit older than that now. I'm getting older than that. And so I think for quite some time, we've had a lot of indications that we really need to remain focused on this.

I do think aerospace and defense continue to inspire and I think the sooner we get back to the Moon and start doing some great things on the Moon, the more we'll be pulling people into these industries, but we need to make sure that they have the basis to come and work.

Diversity also, forgot to mention that, also that's another problem, is that in the last 20–30 years, the face of the workforce really hasn't changed very much and people have worked on it very hard in aerospace and defense and yet it's about the same as it was 40 years ago, yet we know that a more diverse workforce is actually a more competitive work force. So this is another issue that it's a tough–to crack but it's one that we need to keep on working on.

Mr. STALLMER. Yes, I think it's a great question and I see it a little differently.

I see NASA as really one of the marquee government agencies. I just traveled to Europe and everywhere you go, people are wearing NASA logo tee-shirts. NASA is a brand. It is an inspirational brand, and I understand the aging workforce issues.

Why I say I see it a little differently, from the commercial companies that I represent, companies like SpaceX and Blue Origin and Virgin Galactic, they can't hire enough people and there's a line out the door for people that want to work for these type of companies, these innovative, cutting-edge companies that are really pushing the envelope in a fantastic way that's really moving the needle for our industries.

We also represent universities. There are two universities in Arizona that we represent, Arizona State University, where we'll be having our next board meeting, and U of A out there, and the programs that they're having, the innovative programs and studies, the space studies programs that they have are fantastic and they're really preparing these students well to go into the workforce.

I caution myself to tread lightly on this next issue, but there is an immigration problem that we have or maybe a visa problem. We're the most generous country in the world. We educate people from all over the world and we give them one of the finest educations, some of the advanced educations, doctorates in aerospace engineers, and we're so generous that we give them a diploma and we also give them a ticket home.

Those are the people that we want to keep, some of the best and the brightest, instead of sending them back to China and India and elsewhere. Let's keep those people. They can help the work force. As Mary Lynne said, the diversity of the work force, I think we're making efforts. I'm not going to turn around and look, but I know that this week, there's a program called the

Brooke Owens Fellowship Program that gets 36 young women from across the country. Is there anyone from the Brooke Owens Fellowship Program behind us or alumni? I think there are some here.

But they're all coming into town. There are internships all across the country and I think we're doing a better job of inspiring young women to go into the fields of math and science. We're not there yet, but I think the effort is being made and I see a lot of the companies that I work with trying to advance that, as well.

Senator SINEMA. Thank you all so much. Thank you, Mr. Chairman.

Senator CRUZ. Thank you.

Mr. HICKAM. Can I just add? Just as the Chairman of the Board of the Space and Rocket Center, that includes Space Camp and Space Academy. So I would like to add that we have kind of a shadow space workforce program down there. We've trained hundreds of thousands of young people, including 11 astronauts, all women, by the way, so far. So that program down there completely self-funded is doing a remarkable job of getting young men and women interested in STEM.

Senator SINEMA. Thank you. Thank you, Mr. Chairman.

Senator CRUZ. Thank you.

Senator Capito.

Statement of Hon. Shelley Moore Capito, U.S. Senator from West Virginia

Senator CAPITO. Thank you, Mr. Chairman. Thank you to Ranking Member Sinema. It's great to be here to listen and to see my good friend, Mr. Hickam, here, Homer.

You'd be surprised that West Virginia has a great history of space. We started—we didn't start but we do have Chuck Yeager, who was quite a wild one and he's still around. We also had John McBride, who was a commander on the shuttle, and then our mathematician, Katherine Johnson, and then Homer, as well.

I just had the honor of dedicating, renaming the NASA IV&V Center in Fairmont, West Virginia, in honor of Katherine Johnson, and I've heard some very much concern today about the next generation, but I'm going to tell you we had about 250 to 300 people there and by far the greatest majority of them were younger people that were so excited to see and, of course, it's a cross-section of sort of the media and the book but to recognize her and her great accomplishments at NASA over her lifetime.

Roy Lee Cooke was there because we weren't able to get you to come and Roy was waiting for you but, anyway, he's doing well. He's doing well, I report. He was another rocket boy.

The other thing we learned on the Artemis, there was quite a lot of discussion about the Artemis Project because the Administrator was there and did a great job to sort of inform everybody, people don't know about this. This is a problem, I think. Maybe we're not tuned into it, but we're not—I had never heard of Artemis until I walked into that dedication later in the day and it talked about the sustainability of the mission and how we're going to have a long-term presence in and around the Moon.

I said of course we are because it's going to be the first one where a woman is going to go to the Moon and we will sustain it for you guys. So don't worry about that.

Also, Yvonne Cagle was there, who's an astronaut, and she was expressing her desire to go to Mars, which to me sounds just so are you kidding me, go to Mars? You know, I'm having trouble going back and forth from West Virginia to Washington and I said why do you want to go to Mars? Why do you really want to go to Mars? She goes I really want to see what effect it has on my body. She's a physician. So she's very curious about the science of what it would have on her own body and is willing to do this for her but also for the country and for the advancement of science.

In the back of the room, there were robotics teams. There were about 15 of them from high school and middle school and It was interesting because in the robotics teams, they all had mentors who was either a parent or somebody who did something else and did this in their free time once a week, twice a week, and during the competitive times more than that, and I think, you know, about you, Homer, and I think about your book and how the emphasis you had on your teachers in Coalwood and everything, and I thought why aren't we incorporating more of this into the regular curriculum of our schools so that the teachers are there.

Mentors are great, but they're volunteers, and it's tough for them. They have another job. They have other family responsibilities. So I would just like to ask you, Homer, if you could talk about the influence of your teacher and your mentors on you and your investment in science and how you see that now in the context of what we want to do in the future.

Mr. HICKAM. Thank you, Senator Capito, for the question.

Of course, one of the major elements that I wrote about in Rocket Boys was our teachers, especially one teacher, our Ms. Riley, who was our science teacher at Big Creek High School, who so influenced us, who brought us a book that was called Principles of Guided Missile Design when we were trying to figure out how to build rockets.

I later saw that book in a Ph.D. program for rocket science. It required a working knowledge of calculus and differential equations and I personally was having trouble with algebra at the time. But Ms. Riley said all I've done is give you a book. You've got to have the courage to learn what's inside of it.

So she and other inspiring teachers right there in little McDowell County, West Virginia, were pretty amazing and they raised a generation of coal miners' kids who went out and did some pretty great things, I'd have to say, not me especially, but we have captains of industry that came out of that era before, of course, the coal industry ran into problems.

Now the way I look at it now is and try to support that again is that Space and Rocket Center, Space Camp, and Space Academy, Honeywell, Grumman, and other aerospace companies send and pay for teachers to come to Space Camp and I make a special room on my calendar to go out and talk to those teachers when they come in and, my goodness, they are so enthusiastic about the space program. And the neat thing about a teacher, you know, we can teach a young person at Space Camp and Space Academy and they will have all this knowledge, but if you teach a teacher, my goodness, you think of the hundreds and thousands of students that they then are able to pass this knowledge along and the enthusiasm for the space program.

So we must never forget our teachers. God bless them all.

Senator CAPITO. Thanks. Thank you.

Senator CRUZ. Thank you. Senator Gardner.

Statement of Hon. Cory Gardner, U.S. Senator from Colorado

Senator GARDNER. Thank you, Mr. Chairman, and thank you for hosting this hearing today.

This is a pretty incredible panel for a kid who grew up on the Eastern Plains of Colorado and wanted to be an astronaut. I failed miserably at that, but I'll never forget the letters that I wrote to NASA and the pictures I got back from NASA.

Mr. Kranz, Dr. Darden, Mr. Hickam, there's a reason people wrote books about you and made movies about you, that you're featured in our culture and our ideas and our society. It's because of the impact you've had on kids like me.

Turn around and look at the young people in this room. Please turn around. Look at them all. Raise your hand if you've been inspired by the work that any one of these have done on the panel. Please raise your hand right now if you want to lead better lives because of the people on this panel.

[Majority of the room raising their hands.]

Senator GARDNER. This is what it's about. You didn't just change the 1960s. You've changed the world with an impact that will last forever.

I mean, Mr. Hickam, all I've done is given you a book. You have to have the courage to find out what's inside it. That comment that your teacher made is so powerful. Thank you for being here. Thank you for the work you've done, and I hope that as we look at things like the Rising Above the Gathering Storm and the America COMPETES Bill and the American Innovation and Competitiveness Act, the work that we will continue to do to rise above the storm that often is Washington, we'll make every single one of the people that raised their hands in this auditorium, this committee room proud of the policies this Congress pursues and this country pursues because the day that somebody else lands on Mars and not the United States, the day that somebody else invents the next better light bulb, the day that somebody else has created the next better internet, the next better algorithm, is the day that those jobs, those ideas, those teachers, those people go somewhere else, and the power of that innovation, the power of that idea, that thought is no longer here.

We have to maintain that incredible power that is the United States and not the military prowess but the power of the people sitting before this committee, what's in your head, the knowledge that you have that you've given to this country.

Thank you for the work that you continue to do to inspire, to dream, to hope, to aspire, to create, because you've made this country a great place.

Thanks. That's all I have.

Senator CRUZ. Thank you, Senator Gardner, and thank you for in particular pointing out all the young people here and all of us who are inspired.

Dr. Darden, I wanted to ask a follow-up question. As I mentioned in my opening, you and I visited a couple of weeks ago at the dedication of the new street sign in front of the NASA Headquarters, Hidden Figures Way, and so now NASA Headquarters is on a street named after you and named after your colleagues and fellow pioneers and as you and I discussed then, I'm particularly inspired because like you, my mom was a mathematician who became a computer programmer in the 1950s and 1960s, and indeed at the

Smithsonian helped work on computing the orbit of Sputnik and when I went to see the movie Hidden Figures, I took my mom and my wife and both my daughters, and afterwards, I was talking with my girls and telling them that Meme, their grandmother, had been doing much the same as you were doing and your colleagues.

I was commenting—I asked my mother at the time. I said, "All right, how accurate was the movie in terms of what it was like to be a woman and a mathematician in the 1960s?"—or in her case starting in the 1950s. And her reaction was she thought it was quite accurate, and I said one of the odder things to me listening to it was seeing people called computers.

You know, we think of a computer as a hunk of metal on a desk, not a human being who's computing, and my mother began laughing and she said when she came out of Rice in 1956 and went to work at Shell, her first job title was computer and so I would just ask you, Dr. Darden, if you could tell this Committee what it was like to be a human computer, to be helping drive the incredible success of NASA all these years?

Dr. DARDEN. Well, I think we worked with great people. I would say to you, and it's a confession, I did not like being called a computer.

[Laughter.]

Dr. DARDEN. But the support work that we did—I could program when I went there. So the support work that we did with engineers and everything was very great. I think that what really inspired me is to know what the work that I was doing, the equations I worked with, what they were doing to the real world, ——

Senator CRUZ. Right.

Dr. DARDEN.—how they were evaluating the wind tunnels there, such that when we used that data to go fly, everything was correct, and so that really inspired me, but I worked with wonderful people and so that's when I—and I went there as a data analyst, but soon after, about the time that the Apollo Program ended, I began asking to be switched to an engineer and I switched to the engineer at about a year after super-sonic flight across the Continental United States had been banned by law and the SST Program had been canceled, and so that became my life after that and, I mean, I kind of dreamt that and lived with it for 25 years and still hoping that the airplane that Lockheed-Martin builds will be able to have us get rid of that law. We can have supersonic flight in this country.

So it was great. Certainly when Apollo was going up and everything, it was just fantastic to see these things happening in our lives and my vision for the space in a few years, I guess, is that we can actually start operating in space like we operate the airplanes on Earth. So I actually dream of that kind of mobility within space.

Senator CRUZ. I look forward to that.

Mr. Stallmer, in your testimony, you mentioned that "since 2000, investors have supported 375 private space companies with nearly $19 billion in private capital."

Like you, I think that is a wonderful development. I think that is key to expanding into space at the level we need to.

My question for you, and Dr. Dittmar, feel free to chime in here, as well, if you have thoughts, what should we be doing turn that $19 billion to increase it tenfold and then a hundredfold to get the resources invested that will be needed to go to the Moon and build a habitat and go to Mars and go to Europa and explore space?

Mr. STALLMER. I think it's critical for investors to see that the government is a partner, not an adversary, and I wouldn't say that any such agency, you know, is deliberately trying to impede what we are doing, but when you have that partnership, that stability of working with these various government agencies, it provides a sense of assurance, I think, for other investors.

There are a lot of investors out there, not just in the U.S. but around the world, and of that $19 billion, over $4.3 billion of that came from last year alone. So 71 unique investors, you know, that have invested in space companies in the last two years alone. So there's money out there. They want to see the stability of the government.

I think that what the Senate has done with your committee, the bipartisanship of your committee working together has been a real beacon for many to see. I think the President's enthusiasm on the space program with the creation of the National Space Council, the UAG, which Homer and Mary Lynne and I all serve on, and the passion, the reinvigorated passion, I should say, for the Space Program, I think that's sending a great message to investors and the regulatory environment, and I think this was highlighted in the Space Frontier Act.

Now how do we streamline this regulatory environment to make it more accessible and streamlined, you know, to access space? A lot of the rules and regulations that we're under under the current regime go back to the 1980s when there wasn't a commercial space industry at all. It was all government space. So how do we adapt to that, and I think your Committee has done a fantastic job on trying to address these issues, and I thank you both for that.

Senator CRUZ. Thank you.

Dr. Dittmar, did you have——

Dr. DITTMAR. Yes, no. I'll just agree with everything that Eric has said.

I think in my testimony I made some references to essentially how we go forward with government. Government acquisition processes make a lot of sense when you have very long lead time. Programs take a long time. They have a heavy R&D component using traditional acquisition processes under those circumstances make sense, although accountability is also very important and should not be lost.

But I think also the Nation needs to have a conversation with itself about what are the goals with regard to developing an off- Earth economy, however it is you want to talk about it.

Right now, that's sort of been left with NASA, right, because, I mean, it's the space agency, but I think there are reasonable questions to be asked. We need to talk about economic development of low-Earth orbit, for example. NASA is not an economic development agency. It's a space agency.

So it recently rolled out—because Eric's going to talk about this a little bit tomorrow, but NASA recently rolled out some commercial development ideas on July 7th. Does all that belong in NASA's hands? Should we be talking to the Department of Commerce? Does it make sense to sort of get some other—including NGOs, I mean, to sort of look at this?

This is really significant. We're talking about doing economic development off the Earth. I mean, let that sink in for a minute. We're not so hot at it down here. It could take 20 years, it could take 20–30 years, and we don't have the sort of barriers that we have once we're trying to launch.

So I think there needs to be a serious conversation probably in the halls of government, perhaps in Congress and other places, about how do we really go about doing that and how much investment do we want to make in that and what kind of investment? Do we want to start talking about government-backed loans? I mean, there are lots of means to talk about economic development. How do we go about doing it?

Senator CRUZ. Well, thank you very much, and let me thank each of the witnesses for, I think, very helpful and productive testimony. Thank you for being here. Thank you for your careers and lifetime that has inspired so many young girls and young boys and augers well for the future.

The hearing record will remain open for two weeks, closing on July 23. During that time, Senators are asked to submit any questions for the record and upon receipt, the witnesses are requested to submit their written answers to the Committee as soon as possible.

And with that, the hearing is adjourned. [Whereupon, at 4:29 p.m., the hearing was adjourned.]

Chapter 4

NASA's Plan for Sustained Lunar Exploration and Development*

National Aeronautics and Space Administration

Nasa's Plan for a Vibrant Lunar Future

> "Lead an innovative and sustainable program of exploration with commercial and international partners to enable human expansion across the solar system and to bring back to Earth new knowledge and opportunities. Beginning with missions beyond low-Earth orbit, the United States will lead the return of humans to the Moon for long-term exploration and utilization, followed by human missions to Mars and other destinations."
>
> – President Trump, Space Policy Directive 1, December 11, 2017

> "The NASA Administrator shall submit a plan to the Chairman of the National Space Council for sustainable lunar surface exploration and development, including necessary technologies and capabilities to enable initial human exploration of Mars."
>
> – Vice President Pence,6th Public Meeting of the National Space Council, August 20, 2019

* This is an edited, reformatted and augmented version of National Aeronautics and Space Administration Publication.

In: The National Aeronautics and Space Administration's Role …
Editor: Brent M. Jenkins
ISBN: 979-8-89530-539-3

Overview

The Moon is the gateway to the solar system. A world equivalent to an entire continent that human feet have touched only a few times. As Earth's nearest planetary neighbor, the Moon has profound potential to be a source of new scientific advances and economic growth. It is also the best place for us to test our deep space systems and operations in preparation for the first human mission to another planet: Mars. Over the next decade, the Artemis program will lay the foundation for a sustained long-term presence on the lunar surface and use the Moon to validate deep space systems and operations before embarking on the much farther voyage to Mars.

Over the coming decades and generations, our presence will grow to use and develop the extensive resources of the Moon, including its water and metal deposits. As the Moon unveils her secrets, scientific interest continues to grow. In addition to enabling scientific understanding of the formation of the Earth and the solar system through lunar geology and chemistry, exploration of the Moon will enable ground-breaking scientific discoveries about the universe, including through radio astronomy from the incredible vantage point of the far side of the Moon. As in the skies and now low-Earth orbit, NASA's scientific and exploration efforts lead the way and economic development follows. The Artemis program will similarly enable commercial opportunities on the lunar surface, beginning with the first U.S commercial lunar deliveries next year.

The Moon is more than a physical destination. A core focus of Artemis is to extend the nation's geo-strategic and economic sphere to encompass the Moon with international partners and private industry. The United States will build confidence among its commercial, U.S. government, and international partners by leading the development of clear policy principles to support civil space exploration with an initial emphasis on the Artemis program. Specifically, the U.S. will establish a predictable and safe process for the extraction and use of space resources under the auspices of the Outer Space Treaty.

This document covers and responds to the Chairman of the National Space Council's direction to provide a plan for a sustained lunar presence, including the technologies and capabilities to enable the first human mission to Mars. For millennia humanity has looked at the Moon in wonder and awe. As the United States leads the development of a sustained presence on the Moon together with commercial partners and international partners, our presence on the Moon will serve as a constant reminder of the limitless potential of humanity. It will continue to inspire humanity as we seek ever more distant worlds to explore - starting with Mars.

The first human mission to Mars will mark a transformative moment for human civilization. Establishing a sustained lunar presence and taking the initial steps toward the first human mission to Mars will be the greatest feat of engineering, and the greatest voyage of exploration and discovery, in human history. These missions will drive technology and innovation using the country's unparalleled scientific capabilities, dynamic economy, and robust industrial base.

These missions will inspire generations of science, technology, engineering, and mathematics (STEM) professionals and countless other disciplines, while offering opportunities to domestic partners in government, industry, and academia.

Most importantly, the accomplishments of the Moon to Mars approach will assure that America remains at the forefront of exploration and discovery. The United States is still the only nation to have successfully landed humans on the Moon and spacecraft on the surface of Mars. As other nations steadily increase their presence and spending, American leadership is now called for to lead the next phase of humanity's quest to create a future comprised of endless discovery and growth in the final frontier.

The Artemis Generation

Everyone born after the year 2000 has always known a world where people have been living in space. In this decade, we will all see humans walking on the Moon again – this is the Artemis Generation. Artemis, the twin sister of Apollo, will put in place the key infrastructure on and around the Moon that will be built upon and leveraged for generations to come.

Americans will return to the Moon in 2024. Following this 2024 landing, we will develop a sustained, strategic presence at the lunar South Pole called the Artemis Base Camp. Our activities at our Artemis Base Camp over the next decade will pave the way for long-term economic and scientific activity at the Moon, as well as for the first human mission to Mars in the 2030s.

Starting next year, a steady stream of robotic precursors and technology demonstrations will begin lunar operations. The Moon is a fundamental part of our planet's past and future.

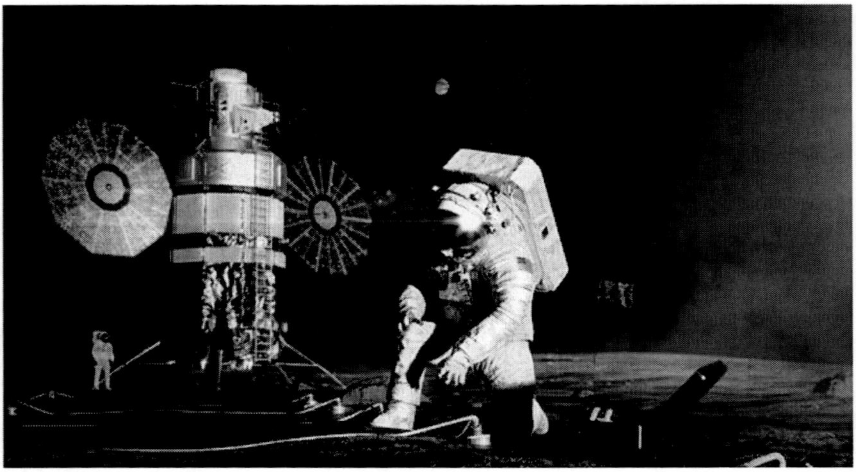

Figure 1. Concept image of an early Artemis expedition on the lunar surface.

Although Americans first walked on its surface more than 50 years ago, our explorers left only fleeting footprints at a few sites, spending a total of 16 days on the lunar surface. These missions were all in the equatorial region, with a total traverse of less than 100kms (~62 miles) – on a body whose surface area is the size of Africa.

This next wave of lunar exploration will be fundamentally different. It starts with American expeditions to the vicinity of the Moon in 2023, and landing astronauts on the surface in 2024. This will be the first chance for most people alive today to witness a human lunar encounter and landing – a moment when, in awe and wonder, the world holds its breath. America will not stop there – this will be the opening chapter of a new era of discovery and exploration.

Nasa's Three Domain Exploration Strategy

Artemis is the core of NASA's exploration and human spaceflight plans for the next decade. Artemis builds upon ongoing human spaceflight efforts conducted aboard the International Space Station (ISS) and prepares the way for future human spaceflight programs, including the first human mission to Mars. NASA's overall Level Zero Goals for exploration encompass these three primary domains – low-Earth orbit, the Moon, and Mars.

NASA Exploration Level Zero Goals

- Transition U.S. human spaceflight in low-Earth orbit to commercial operations, which support NASA and its partners. Use the ISS and new commercial facilities as testbeds for exploration technologies and to nurture emerging commercial applications.
- Advance long-term robotic exploration of the Moon with robust commercial and international partnerships.
- Land American astronauts on the Moon and return them safely.
- Expand U.S. human spaceflight operations at the Moon to support sustained lunar surface activities and to demonstrate elements of a Mars-forward architecture.
- Continue U.S. leadership at Mars by advancing robotic access in preparation for human exploration.
- Engage and inspire America and the world along each step of the way.

Figure 2. The transition to low-Earth orbit has already begun. Two commercial cargo vehicles, SpaceX's Dragon and Northrop Grumman's Cygnus, service the International Space Station.

Pursuing these goals will ensure our current and future activities in low-Earth orbit can enable Artemis, and that Artemis enables the future exploration of Mars.

NASA has developed a strategy for achieving the Level Zero Goals that includes: a combination of robotic and human missions; architectures that build off of key, enabling hardware elements already in development; new technologies that will expand future options and operations; and new developments which will engage the NASA workforce, industry, and international partners to encourage innovative capabilities and harness competitive and cooperative energies. These efforts are being integrated into a sequence of missions that start with a near-term return of humans to the Moon, and then continue with a set of missions on and around the Moon that will lay the foundation for a sustainable presence. Ultimately, the experiences gained, and technologies demonstrated by these lunar operations will support a historic first human mission to Mars. Conducting scientific exploration synergistically with crew and robotic explorers teaches us effective techniques that can be applied as we push the boundaries of space exploration.

Robotic Missions

Our return to the Moon begins with robots. For more than a decade, NASA's Lunar Reconnaissance Orbiter has been imaging and mapping the Moon for scientific research and in preparation for a human return. NASA's new Commercial Lunar Payload Services (CLPS) initiative has already selected the first two robotic missions that, beginning in 2021, will deliver science and commercial payloads to the surface of the Moon. NASA has also committed to using this approach to deliver its next robotic lunar rover, the Volatiles Investigating Polar Exploration Rover (VIPER), which will conduct science investigations of the lunar volatiles at the Moon's South Pole. The data produced by VIPER will inform future in-situ resource utilization (ISRU) technologies. On the lunar surface we will demonstrate precision landing, starting with CLPS deliveries to build experience and improve capabilities that will enhance all future landings, human and robotic. Later, the CLPS approach will be used to deliver other large cargo elements in direct support of human lunar missions.

Overall, while orbital missions have provided extensive information about the lunar surface and its potential resources, robotic lunar surface scouts are essential to validate these observations and prepare for human habitation and utilization of the Moon's rich array of resources from volatiles to minerals. Landers and rovers provide excellent platforms to demonstrate technologies that will enable greater lunar surface mission capabilities and have applications that extend beyond the Moon to Mars initiative, such as terrestrial robotic mining systems and next-generation power storage. Multiple landers will provide a global view of the Moon and its resources. Rovers will be used to explore the surface more extensively, carrying a variety of instruments including ISRU experiments that will generate detailed information on the availability and extraction of usable resources (e.g., oxygen and water).

These robotic efforts will unleash a broad array of inquiry and scientific investigations. The Moon is a natural laboratory to study planetary processes and evolution, and a platform from which to observe the universe. Bombarded by solar and cosmic radiation for billions of years and left largely undisturbed, the Moon is a historic archive of our Sun and solar system. Scientific discoveries are locked in its regolith that could lead to improved understanding of our own planet and its evolution. The far side of the Moon offers an unparalled window to look back into the beginning of the universe. It also harbors resources, such as water, that are among the rarest and most precious commodities in space, offering potential sustenance and fuel for future explorers.

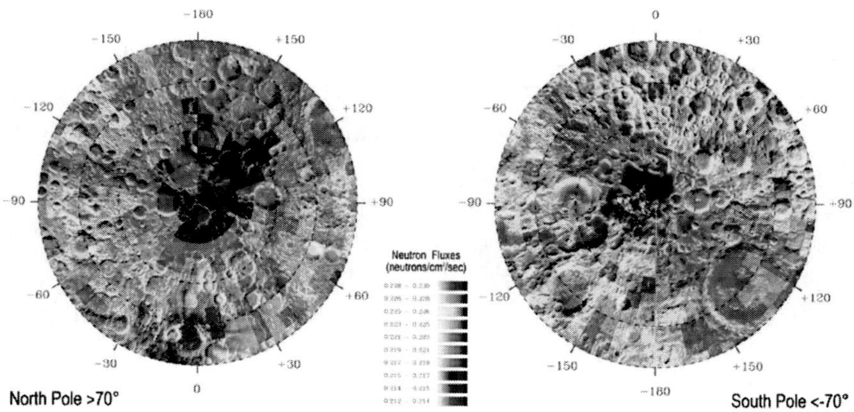

Figure 3. Image Above: The dark blue and purple areas at the Moon's poles indicate neutron emissions that are consistent with hydrogen-rich deposits covered by desiccated regolith. These hydrogen signatures are possible indications of water in the form of water or hydrated minerals. Feldman et al., Science, 281, 1496, 1998.

The Early Artemis Missions

The foundation for our return to the Moon is NASA's Orion spacecraft and Space Launch System (SLS). The Orion spacecraft has been designed for deep space operations around the Moon for up to four crew members, and the SLS is the powerful heavy-lift rocket designed to launch it, and potentially other high-mass cargo to the lunar environment. Added to these are the two newest elements of the lunar return architecture under contract, the power and propulsion element (PPE) and the habitation and logistics outpost HALO). Together they form the Gateway's foundation as the Artemis lunar orbiting platform. Early Gateway operations will be autonomous, with help from NASA's Mission Control Center at Johnson Space Center in Houston to conduct systems checkouts and capture critical scientific data about the deep space environment. Orion will deliver the first crew to Gateway when the human landing system (HLS) capability can enable lunar expeditions to be staged from the stable Gateway orbit. Building confidence in this system of an orbiting command module and deployable landers will serve as a critical analog for human missions to the surface of Mars. It will also serve as a strategic capability – allowing access and presence in the orbital lanes around the Moon and to the rest of the solar system.

The Artemis program begins with an uncrewed flight test of the SLS and Orion (Artemis I), then a crewed flight test (Artemis II). Artemis I will see SLS send an uncrewed Orion 280,000 miles from Earth, thousands of miles beyond the Moon over the course of an approximately three- week mission. Mission controllers on Earth will collect data to assess the performance of both spacecraft. This mission will also deploy 13 CubeSats that will conduct new scientific investigations and new technology demonstrations that will engage a broader set of universities and companies in lunar exploration than ever before on a single mission.

With the first crewed flight test of the SLS and Orion, Artemis II, astronauts will return to the vicinity of the Moon for the first time in more than 50 years. This will be an Apollo 8 moment for a new generation. At the end of this mission, NASA intends to have tested every hardware, software, and operational component of Artemis III except for the actual landing on the surface.

Artemis III will be the culmination of the rigorous testing and nearly one million miles of flight demonstrations on the deep space transportation systems that NASA will accumulate during Artemis I and II. When Artemis III lands the first woman and next man on the Moon in 2024, America will have demonstrated a new level of global space leadership. With this robust lunar exploration capability re-established, NASA and the world will focus on building a sustained presence on the lunar surface in preparation for long-term development on the Moon and the human exploration of Mars.

Figure 4. Concept image of the Space Launch System launching from Kennedy Space Center.

The Gateway will establish U.S. leadership and a sustained presence in the region between the Moon and Earth. The platform will offer astronauts easier crew returns, a safe haven in the event of an emergency, the ability to navigate to different orbits around the Moon and later, an advancement in human life support systems.

Gateway will expand to include critical contributions from international partners, specifically, a robotic arm, substantial additional habitation volume, and refueling capabilities. Canada announced in February 2019 that it intends to participate in the Gateway and contribute advanced external robotics. In October 2019, Japan announced plans to join the United States on the Gateway with contributions to habitation components and logistics resupply. In November 2019, the European Space Agency received authorization and funding to support its planned contributions to the Gateway, the International Habitat (I-Hab), and the European System Providing Refueling Infrastructure and Telecommunications (ESPRIT), both of which will dramatically enhance the capabilities of Gateway, contributing to sustainable operations while paving the way for a future human mission to Mars. Russia has also expressed interest in cooperating on the Gateway via the contribution of an airlock. The Gateway will provide a next- generation deep space platform from which to conduct science investigations outside the protection of the Earth's Van Allen radiation belts. The international science community has identified heliophysics, radiation, and space weather as high-priority investigations to conduct on the Gateway. The first two Gateway payloads are a radiation instrument package provided by the European Space Agency and a space weather instrument from NASA. The agency also recently awarded the first Gateway Logistics Services (GLS) contract to SpaceX to deliver cargo, experiment and other supplies to the outpost. Echoing the success of the Commercial Resupply Services program, GLS will leverage commercial partners to deliver logistics to the Gateway, supporting lunar operations while building experience and technologies for future logistics missions that can support the first human mission to Mars.

With these core elements, logistics support, and flights of SLS/Orion underway, and the acquisition of HLS in progress, NASA is opening up other core elements of a sustained lunar presence – including the lunar terrain vehicle (LTV), the lunar mobile habitat or habitable mobility platform, the lunar foundation surface habitat (FSH), power systems, lunar ISRU systems, and expanded Gateway habitation capabilities – with new international and industry partnerships. With this approach, NASA will leverage years of hard work and national investment in the systems needed to return to the Moon, while enabling and using new partners and new capabilities to ensure that our return to the Moon is sustainable and leads directly to the first human mission to Mars.

Artemis After 2024

After Artemis III, the overall plan is to conduct operations on and around the Moon that help prepare us for the mission durations and activities that we will experience during the first human mission to Mars, while also emplacing and building the infrastructure, systems, and robotic missions that can enable a sustained lunar surface presence. To do this, we will develop Artemis Base Camp at the South Pole of the Moon.

Artemis Base Camp will be our first sustainable foothold on the lunar frontier. We will initially move to one to two-month stays to learn more about the Moon and the universe. We will develop new technologies that advance our national industries and discover new resources that will help grow our economy. Overall, the base camp will demonstrate America's continued leadership in space and prepare us to undertake humanity's first mission to Mars.

The three primary mission elements of Artemis Base Camp are: The LTV that can transport crew around the site; the habitable mobility platform for long- duration trips away from Artemis Base Camp and the foundation surface habitat will enable short-stays for four crew on the lunar South Pole. Combined with supporting infrastructure added over time such as communications, power, radiation shielding, a landing pad, waste disposal, and storage planning – these elements comprise a sustained capability on the Moon that can be revisited and built upon over the coming decades.

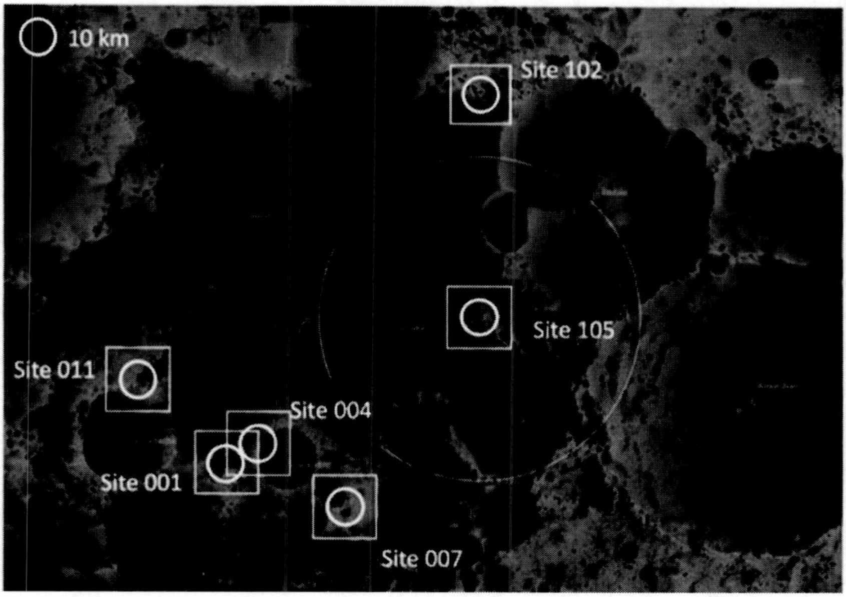

Figure 5. A South Pole landing site has not been determined, but this image shows sites of interest near permanently shadowed regions, which may contain mission-enhancing volatiles. These sites may also offer long-duration access to sunlight, direct- to-Earth communication, surface slope and roughness that will be less challenging for landers and astronauts.

Mobility is a major part of the Artemis Base Camp. The LTV and the habitable mobility platform will enable long-term exploration and development of the Moon. In addition to its size, the Moon's geography is complex, and its resources dispersed. Looking at potential sites for Artemis Base Camp, such as near Shackleton Crater, shows the immense scale of the lunar geography. Robust mobility systems will be needed to explore and develop the Moon. The same is true for Mars, making the habitable mobility platform a particularly important element as we will need a similar type of vehicle to explore the Red Planet.

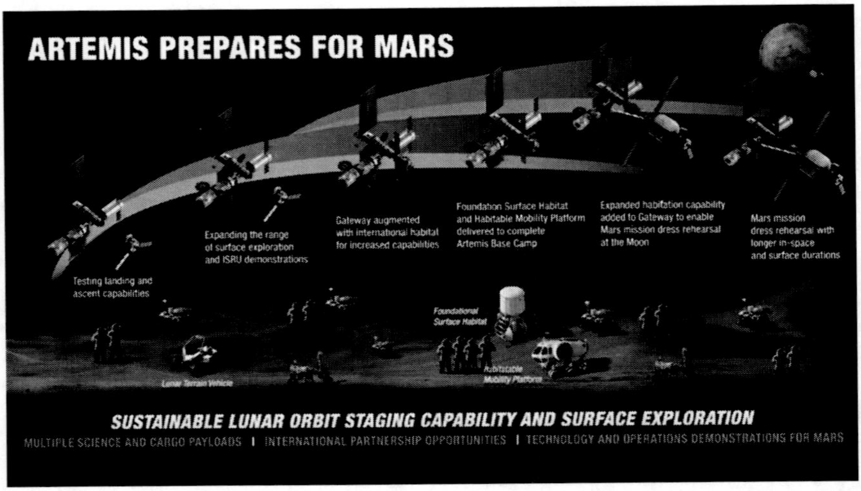

Figure 6. Artemis Base Camp evolves on the surface as the Gateway is leveraged for Mars preparation.

In addition to establishing Artemis Base Camp, another core element of the sustained lunar presence that feeds forward to Mars will be the expansion of habitation and related support systems at the Gateway. This evolution of the Gateway's systems to include large-volume deep space habitation would allow our astronauts to test, initially in lunar orbit, how they will live on their voyage to and from Mars. Gateway can also support our first Mars mission analogs on the lunar surface. For such a mission, we currently envision a four-person crew traveling to the Gateway and living aboard the outpost for a multi-month stay to simulate the outbound trip to Mars, followed by two crew travelling down to and exploring the lunar surface with the habitable mobility platform, while the remaining two crew stay aboard. The four crew are then reunited at the Gateway for another multi-month stay, simulating the return trip to Earth, before landing back home. These missions will be by far the longest duration human deep space missions in history. They will be the first operational tests of the readiness of our long-duration deep space systems, and of the split crew operations that are vital to our approach for the first human Mars mission.

Figure 7. The lunar South Pole's Shackleton Crater, as captured by the Lunar Reconnaissance Orbiter, with the Capital Beltway overlaid for scale.

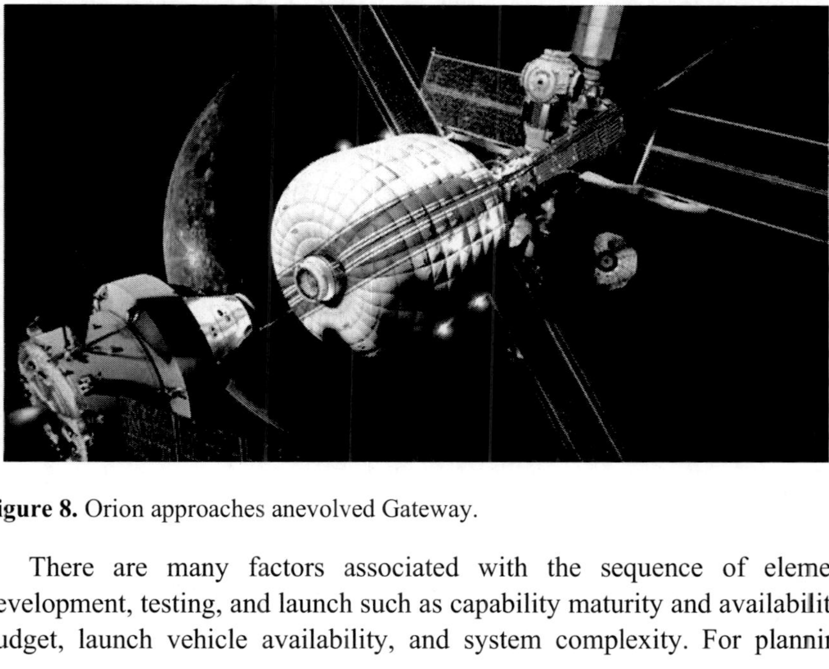

Figure 8. Orion approaches anevolved Gateway.

There are many factors associated with the sequence of element development, testing, and launch such as capability maturity and availability, budget, launch vehicle availability, and system complexity. For planning purposes, NASA is developing a sequence that accounts for these variables and results in an annual cadence of demonstrable progress and a gradual increase in mission duration and complexity. This plan results in the development and emplacement of the infrastructure required for a long-term sustained lunar surface presence while testing systems and gaining the operational experience required for the human Mars mission.

The sequence as currently envisioned begins by sending lunar precursor robotic missions including VIPER by CLPS landers to provide ground truth of terrain, as well as water and metal resource availability for the human lunar landing site. To provide mobility and extended range of exploration for the first several human lunar surface missions, the LTV will be delivered to the lunar surface. The first elements of the lunar Gateway are in development and will support later sustainable human lunar landing missions. NASA anticipates its international partners will provide at a minimum the robotic arm, I-Hab, and ESPRIT to supplement the Gateway's capabilities in lunar orbit.

The habitable mobility platform will be delivered to the lunar surface to expand our exploration range by tens of kilometers and mission duration on the surface from 7 days to 30-45 days, enabling potential Mars surface analog missions on the lunar surface. Other key pieces of the Artemis Base Camp infrastructure are also delivered, including the foundation surface habitat, which will support a crew up to four on the lunar surface, the lunar surface power systems, ISRU demonstrations and pilot plants.

An evolved Gateway habitation capability in lunar orbit will allow us to begin the methodical lengthening of mission durations. This approach will also allow NASA to test risk mitigation approaches for long-duration mission crew and element systems risks that are required for two- year Mars class missions.

Once these pieces of the Moon to Mars campaign are delivered and operational, annual human missions with increasingly long durations will enhance the exploration and sustainable development of the lunar surface.

A Vibrant Earth-Moon Future

Whenever the first human mission to Mars occurs, it will not mean that we are done with the Moon. The windows for launching the two-year mission to Mars open up every few years, and we will continue to conduct human missions to the lunar surface to test systems, conduct scientific investigations, and continue to develop our sustainable lunar presence as we prepare for the optimal launch window.

We will continue to explore the Moon indefinitely -- leveraging robotic deliveries provided by CLPS providers, longer duration human missions, and commercial and international partnerships that will add to the Artemis Base Camp elements NASA puts in place. The LTV, habitable mobility platform and foundation surface habitat will stay on the Moon enabling crews to live on the lunar surface for months at a time. The rovers and crew-tended capabilities on the Moon will be designed to operate autonomously and to work with independent robotic assistants.

In addition to testing our systems for the first mission to Mars, a core purpose of Artemis Base Camp will be to demonstrate new technologies that, over time, will expand our capabilities and reduce the costs of lunar operations. Astronauts at Artemis Base Camp will be testing a wide set of new technologies in six priority areas encompassed by the recently announced Lunar Surface Innovation Initiative (i.e. ISRU; surface power; extreme access; excavation and construction; lunar dust mitigation; and extreme environments). Some of these technologies will help from the beginning, such as lunar dust mitigation and enabling operations in extreme environments like the cold of the lunar night. Other technologies are in early development for significant long-term benefits. For example, ISRU will enable the production of fuel, water, and/or oxygen from local materials, enabling sustainable surface operations with decreasing supply needs from Earth. For surface power, our goal is to develop advanced solar collection and a small, lightweight fission power system to support even longer-duration missions and operations on the Moon, and eventually for Mars and beyond. Autonomous manufacturing, excavation, and construction technologies will make infrastructure emplacement more affordable.

Astronauts will also conduct tests of advanced robotics, including future biomimetic systems that enable more autonomous operations at the Moon and can serve as robotic assistants to the crew. In time, Artemis Base Camp might also include a hopper that could deliver science and technology payloads all over the Moon and which could be operated by crew at Artemis Base Camp and refueled using locally sourced propellant. A lunar far-side radio telescope could also be remotely emplaced and operated from Artemis Base Camp – a sort of backyard radio- telescope at our first encampment on the Moon.

Developing a sustained and vibrant lunar presence will not only require the best of NASA and its international partner space agencies, but also the best of the entire U.S. government and commercial sector. Establishing the infrastructure that will enable additional international and commercial partnerships will result in opportunities for the first international and commercial astronauts on the lunar surface, opening up Earth's 's off-shore continent for ongoing human discovery and development. NASA's unique leadership and capabilities will be combined with innovation and contributions from the same sectors that fuel our nation and economy here on Earth.

The Initial Human Mars Mission

The success of the first human mission to Mars requires a voyage of stunning technological and operational complexity. The distance from Earth to the Moon is a relatively manageable 250,000 miles. In contrast, Mars is, on average, 140 million miles from Earth. The challenges of a mission to Mars are compounded not only by these distances but by the more dangerous levels and types of radiation that is found in deep space. Mars' atmosphere – while intriguing from a scientific perspective – also presents distinct challenges for getting humans to and from its surface.

The initial lunar missions will greatly inform our concept of operations for Mars. Every day in deep space increases the probability of catastrophic events occurring. The concept of operations NASA is working toward for the first human mission to Mars is therefore one that reduces trip time significantly and minimizes time spent on the surface to around 30-45 days. Factors NASA will continue to consider include: risks to crew health during transit, both from galactic cosmic ray (GCR) radiation and from potential catastrophic mission events; the complexity of mission operations on the Martian surface; and the complexity and cost of mission systems, such as in-space propulsion, the ability to land heavy payloads, and required surface systems. NASA has targeted conducting a human mission to Mars that can be realized as soon as possible, while still ensuring that our Mars surface capabilities will allow for extensive exploration on our first mission, including a search for Martian life. The Moon will allow us to test and demonstrate significant parts of this mission before sending humans on this epic journey. NASA will have more to share on the strategy for how the Moon prepares us for the first human mission to Mars in the coming months.

Artemis and the development of Artemis Base Camp will inspire the world with the ability and commitment of American leadership, and in the positive potential of humanity as a whole. If we are to leave a legacy of greatness, hope, limitless opportunity, and growth to future generations, then it is a mission we cannot afford to postpone.

Chapter 5

Current IV&V Projects*

National Aeronautics and Space Administration

Current IV&V Projects

NASA's IV&V Program is currently performing IV&V for the projects listed below

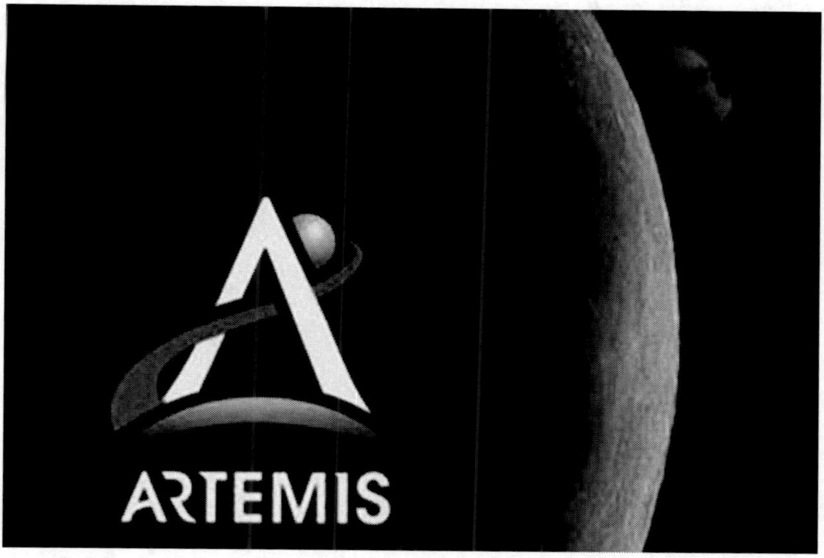

The symbol for the Artemis Program is a blue Earth with an A on top and a red swoosh that goes past the Moon toward Mars.

* This is an edited, reformatted and augmented version of National Aeronautics and Space Administration Publication.

In: The National Aeronautics and Space Administration's Role …
Editor: Brent M. Jenkins
ISBN: 979-8-89530-539-3

Artemis

With NASA's Artemis campaign, we are exploring the Moon for scientific discovery, technology advancement, and to learn how to live and work on another world as we prepare for human missions to Mars. We will collaborate with commercial and international partners and establish the first long-term presence on the Moon. NASA will land the first woman and first person of color on the Moon, using innovative technologies to explore more of the lunar surface than ever before.

Concept art of Dragonfly drone.

Dragonfly

Dragonfly is a drone-like rotorcraft that would explore the prebiotic chemistry and habitability of dozens of sites on Saturn's moon Titan, an ocean world in our solar system. Dragonfly will study dozens of promising sites around Saturn's icy moon Titan and advance our search for the building blocks of life in the universe.

Concept art of NASA's Europa Clipper.

Europa Clipper

NASA's planned mission would conduct detailed reconnaissance of Jupiter's moon and investigate whether the icy moon could harbor conditions suitable for life.

Aerial view of teams moving the Space Launch System core stage into the Vehicle Assembly Building at Kennedy Space Center following its arrival.

Exploration Ground Systems

Exploration Ground Systems' primary objective is to prepare the center to process and launch the next-generation vehicles and spacecraft designed to achieve NASA's goals for space exploration.

Illustration of Artemis astronauts on the Moon.

Extravehicular Activity and Human Surface Mobility

NASA's Extravehicular Activity and Human Surface Mobility Program provides safe, reliable, and effective spacewalking and surface mobility capabilities that allow astronauts to survive and work outside the confines of a base spacecraft in order to explore on and around the Moon.

An artist's render shows the Gateway lunar space station in near rectilinear halo orbit around the Moon. Credit: NASA.

Gateway

Gateway is central to the NASA-led Artemis missions to return to the Moon for scientific discovery and chart a path for the first human missions to Mars and beyond. The small space station will be a multi-purpose outpost supporting lunar surface missions, science in lunar orbit, and human exploration further into the cosmos. NASA is working with commercial and international partners to build humanity's Gateway.

Artist's concept of SpaceX Starship human landing system.

Human Landing System (HLS)

The Human Landing System (HLS) is the mode of transportation that will take astronauts to the lunar surface as part of the Artemis program, including the first woman and the first person of color.

The International Space Station is pictured from the SpaceX Crew Dragon Endeavour lunar space station in near rectilinear halo orbit around the Moon. Credit: NASA.

International Space Station (ISS)

The International Space Station is an unprecedented achievement in global human endeavors to conceive, plan, build, operate, and utilize a research platform in space. With the assembly of the space station at its completion and the support of a full-time crew of six, a new era of utilization for research has begun.

An artist's concept depicts a Mars ascent vehicle starting a sample of Mars rocks on their trip to Earth.

Mars Sample Return (MSR)

Mars Sample Return (MSR) campaign will bring pristine samples from Mars to Earth for scientific study. Once on Mars, Perseverance aims to cache rock and regolith samples in its collection tubes. It then would leave some of them on the Martian surface for an ESA-provided "fetch" rover to collect and deliver to a NASA-provided Mars Ascent Vehicle, which then would launch the samples into orbit around Mars. An ESA-provided Earth Return Orbiter would then rendezvous with the samples in orbit around Mars and take them in a highly secure containment capsule for return to Earth in the 2030s.

A visualization of the Orion spacecraft orbiting the Moon while utilizing a new cutting-edge technology, the Orion Artemis II Communications System (O2O).

Orion

NASA's Orion spacecraft is built to take humans farther than they've ever gone before. On Artemis missions, Orion will carry the crew to space, provide emergency abort capability, sustain the crew during the space travel, and provide safe re-entry from deep space return velocities.

NASA's Space Launch System (SLS) rocket with the Orion spacecraft aboard is seen atop the mobile launcher.

Space Launch System (SLS)

The Space Launch System (SLS) Program will develop a heavy-lift launch vehicle to expand human presence to celestial destinations beyond low Earth orbit. This launch vehicle will be capable of lifting the Orion MPCV to asteroids, the moon, Lagrange points and, ultimately, to Mars.

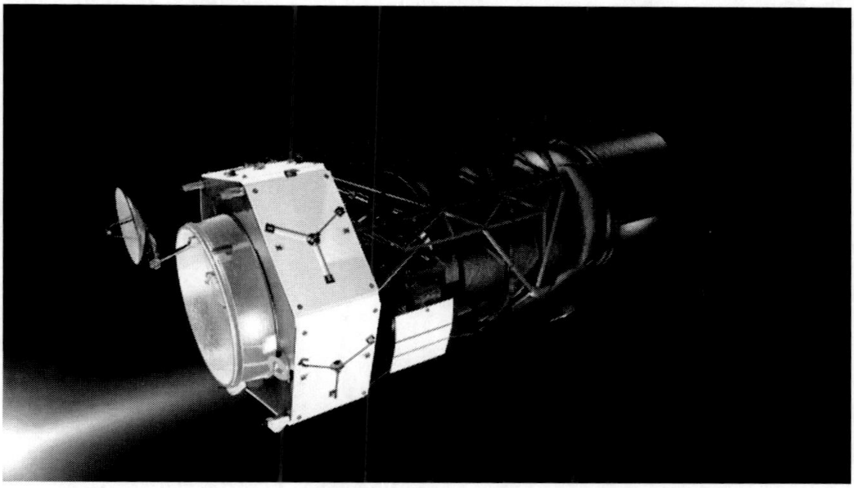

An artist's concept of Roman Space Telescope (Formerly WFIRST)

Roman Space Telescope (Formerly WFIRST)

The Roman Space Telescope (formerly known as WFIRST, the Wide Field Infrared Survey Telescope) – a mission concept to answer vital questions in both exoplanet detection and dark energy research.

Chapter 6

NASA's Deep Space Exploration Plans *

National Aeronauics and Space Administration

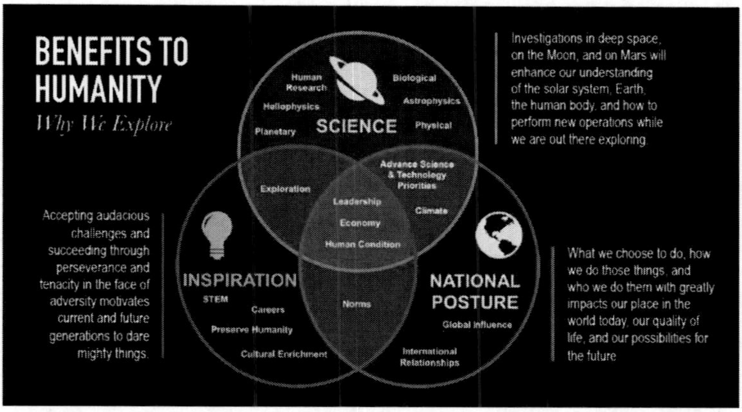

The three pillars of rationale for exploration: science, inspiration, and national posture.

Exploring the Secrets of the Moon and Beyond for the Benefit of Humanity

The Artemis era is here, heralding a future of long-term human exploration in deep space. Through Artemis, the United States will lead an international and commercial coalition to advance scientific research, maintain our national posture, and inspire a new generation of explorers: the Artemis Generation.

While maintaining American leadership in human and robotic lunar exploration NASA will build a global alliance toexplore deep space as one.

---

* This is an edited, reformatted and augmented version of National Aeronauics and Space Administration's Deep Space Exploration Plans Publication.

In: The National Aeronautics and Space Administration's Role …
Editor: Brent M. Jenkins
ISBN: 979-8-89530-539-3

NASA and its partners will explore more of the Moon than ever before with highly trainedastronauts and advanced robotics. We will collaborate with our partners to establish the first long-term presence on the Moon, learn how to live and work away from home, and prepare for future human exploration of the Red Planet. This is America's Moon to Mars exploration approach. Ultimately, what we build, test, and discover at the Moon willdetermine the human capacity to live and work in deep space for longer periods of time, setting NASA on a steady path to send the first astronauts to Mars.

Moon to Mars Architecture

To successfully achieve a lasting human presence in deep space, NASA strategically prioritizes hardware development by beginning with the agency's widely recognized blueprint for exploration and its supporting Moon to Mars objectives that were developed with input from experts across the world. Each objective is decomposed through a systems engineering process to arrive at architectural elements, like rockets, spacecraft, rovers, spacesuits, communications relays, and more—that will be incrementally developed and delivered to the Moon and Mars for long-term, human- led scientific discovery in deep space.

The architecture itself is organized by segments that allow NASA to break the architecture down into manageable pieces to focus and prioritize its analysis work and coordinate with partners. The architecture segments—Human Lunar Return, Foundational Exploration, Sustained Lunar Evolution, and Humans to Mars—are described below.

Architecture Segments

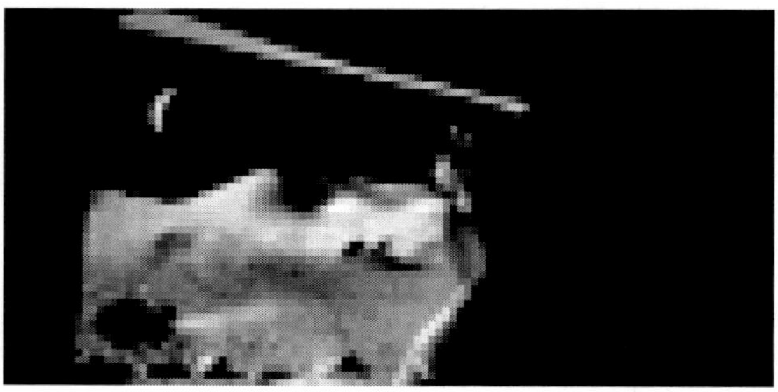

Human Lunar Return

Initial capabilities, systems, and operations necessary to re-establish human presence and initial utilization (e.g., science) on and around the Moon.

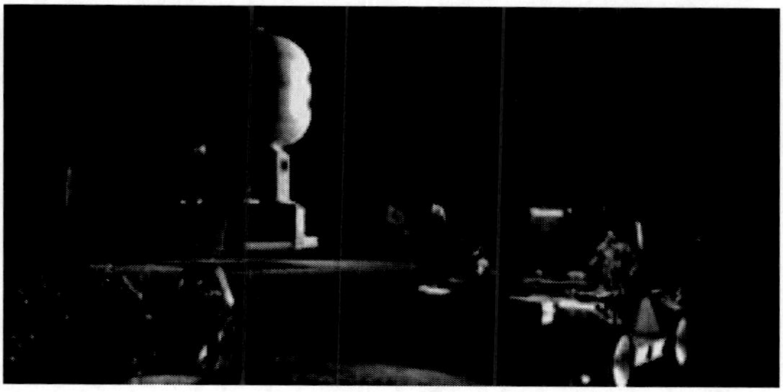

Foundational Exploration

Expansion of lunar capabilities, systems, and operationssupporting complex orbital and surface missions to conduct utilization (e.g., science) and Mars-forward precursor missions.

Sustained Lunar Evolution

Enabling capabilities, systems, and operations to support regional and global utilization (e.g., science), economic opportunity, and a steady cadence of human presence on and around the Moon.

Humans to Mars

Initial capabilities, systems, and operations necessary to establish human presence and initial utilization (e.g., science) on Mars and continued exploration.

Human Lunar Return Elements

Space Launch System

The Space Launch System (SLS) is an evolvable rocket that will provide heavy-lift capability to deliver Orion and its crew, supplies, and other payloads safely to the Moon. In 2022, NASA's Artemis I flight test successfully demonstrated the rocket's capability to send Orion on a lunar trajectory.

Future evolutions of SLS will feature a more powerful upper stage and larger payload volume to send large elements to the Moon, along with Orion and its crew.

Orion Spacecraft

Orion transports up to four astronauts between Earth and lunar orbit and sustains them for up to 21 days. It also provides supplemental power and life-support systems to Gateway while docked to the lunar space station.

Orion launches from Earth atop the SLS, as it did in the Artemis I flight test, and includes the crew module, service module, and launch abort system.

Exploration Ground Systems

Exploration Ground Systems (EGS) provides the ground infrastructure to process and launch SLS and Orion, and also provides recovery capabilities when Orion returns to Earth. EGS uses the Vehicle Assembly Building (shown left) to stack and integrate SLS with its payloads, including Orion.

The Mobile Launcher is another key EGS capability that secures the SLS stack for transportation to Launch Pad 39B atop the crawler-transporter. The Mobile Launcher will be upgraded in future segments to accommodate the evolved SLS.

Gateway

The Gateway ncludes several modules that are incrementally launched and assembled in lunar orbit. In the Human Lunar Return segment, the Power and Propulsion Element (PPE) and Habitation and Logistics Outpost (HALO) are launched together, along with three science instrument suites on a commercial rocket to the near-rectilinear halo orbit.

I-HAB, the International Habitat provided ESA (European Space Agency) and JAXA (Japan Aerospace Exploration Agency) then joinsPPE and HALO for additional habitation volume. This segment alsosees the first Gateway Logistics Module delivery with supplies to outfitthe station. ESA, JAXA, and the Canadian Space Agency will all provide Gateway elements for later segments of the lunar architecture.

Deep Space Logistics

Deep Space Logistics leads the commercial deep space supply chain to transport cargo, payloads, equipment, and consumables to enable exploration of the Moon and Mars. Logistics flights supply Gateway with critical cargo deliveries to maximize the length of crew stays aboard Gateway.

At least one logistics delivery is anticipated for each 30-day Artemis expedition aboard Gateway, and additional capabilities may be added in future segments. An illustration of a commercial logistics module is shown in the image to the left.

Human Landing System

The Human Landing System (HLS) will transport astronauts and their cargo between lunar orbit and the surface. Crew will board the HLS either from Orion or from Gateway.

In this segment, HLS provides the habitable volume, consumables, and design features to allow crew to conduct moonwalks. Later segments will include additional cargo delivery options and habitation on the surface. A commercial provider has been selected for the Artemis III and Artemis IV missions (illustration shown, left). NASA will work with multiple providers to procure landers, and, eventually, landing services for Artemis V and beyond.

xEVA System

The Exploration Extravehicular Activity System (xEVA) includes the lunar surface spacesuit, tools, and vehicle interface equipment that crew will need to conduct moonwalks, collect samples, and deploy science instruments and technology demonstrations.

NASA has two spacesuit providers on contract who can bid to provide spacesuits as services to the International Space Station or to Artemis missions.

Comm, Positioning, Navigation, Timing

Communication, Positioning, Navigation, and Timing (CPNT) services are provided through a combination of assets on Earth, in lunar orbit, and on the lunar surface. On Earth, upgrades to the Deep Space Network, and global installation of the Lunar Exploration Ground System, will provide continuous coverage of the Moon with support from systems in lunar orbit, including Gateway and the Lunar Communications Relay and Navigation System.

CPNT capabilities will grow throughout this segment and beyond, with opportunity for commercial and international partners to provide enhanced or augmented services.

Commercial Lunar Payload Services

Through the Commercial Lunar Payload Services (CLPS) initiative, NASA is acquiring services from a provider pool of U.S. companies to deliver payloads to the lunar surface.

Although some equipment will be delivered to the surface with the crew aboard HLS, CLPS provides an additional capability to deliver cargo and science instruments with commercial providers, supporting a growing lunar economy. NASA's Science Mission Directorate currently has more than 40 U.S. and international science and tech- nology payloads scheduled to be delivered through CLPS providers.

The evolutionary architecture development process described earlier is methodical and deliberate. NASA's intention is to foster transparency so that current and future partners can engage in productive discussions about future collaborations via available, open-source information and shared expectations and objectives.

To reach the required level of maturity for segments beyond Human Lunar Return, the process must continue. Through subsequent analysis cycles and annual Architecture Concept Reviews, the Moon to Mars architecture will grow more defined over time. Implementation of the architecture has already begun for the Human Lunar Return, with elements in that segment directly traced to Moon to Mars Objectives. Objectives distillation for Foundational Exploration and Sustained Lunar Evolution has also begun, as shown in Appendix A of the Architecture Definition Document.

The Humans to Mars segment studies center around early human missions to Mars to define key areas that will inform long-term investment strategies for the larger architectural components, many of which can find early or prototype development in the lunar segments. The ability to identify common systems and operations for both the Moon and Mars provides an additional level of confidence for the earliest human missions to the Red Planet. Decisions made about the Humans to Mars segment will be updated in future versions of the Architecture Definition Document.

Foundational Exploration

The *Foundational Exploration* segment continues using capabilities established in *Human Lunar Return* and initiates new systems to expand mission durations and surface exploration range, increasing the diversity of scientific sampling and research and prepare for human missions to Mars.

Additions to Gateway and the delivery of habitation and mobility elements on the surface will introduce longer mission timelines, accommodating more opportunities to conduct an array science and demonstrate new technologies. The extended mission durations and co-located assets in orbit and on the surface will help optimize operations to conduct Mars mission simulations. *Foundational Exploration* also will be the starting point for activities and capabilities featured in the *Sustainable Lunar Evolution* segment.

NASA used the same process applied in *Human Lunar Return* to distill Moon to Mars Objectives for *Foundational Exploration*, but work remains to map the Functions and Use Cases to specific elements. The general functional capabilities under analysis now for the lunar surface include unpressurized mobility, pressurized mobility, surface habitation, and cargo transportation. In orbit, Gateway's capabilities will grow with contributions from international partners.

Sustained Lunar Evolution

In the *Sustained Lunar Evolution* segment, NASA aims to build, togetherwith its partners, a future of economic opportunity, expanded scientific discovery, and greater participation on and around the Moon.

This "open canvas" segment embraces new ideas, systems, and partners to grow beyond *Foundational Exploration*, increasing globallunar science capability, making longer stays for more astronauts and researchers possible, and leading to the large-scale production of goods and services derived from lunar resources.

It is premature to determine specific elements for *Sustained Lunar Evolution* beyond the higher-level capabilities associated with the Moon to Mars Objectives. For this segment, the focus is on expanding power generation and storage; using lunar resources for propellant; crew consumables; construction materials; and expanding mobility and habitation to accommodate larger populations on the lunar surface.

Humans to Mars

Like the previous two segments, the Use Cases and Functions for Humans to Mars have not been fully distilled from the Moon to Mars Objectives to arrive at the specific elements needed for the first human missions to Mars.

To support ongoing analysis for Humans to Mars, NASA is studying a range of potential mission assumptions to inform system and technology concepts. To illustrate the range of potential Mars architecture solutions, some representative initial assumptions for the current body of analysis work include, at a minimum:

- A light initial exploration footprint: four crew members to Mars orbit with two crew members descending and living on the surface for a 30-sol surface stay;
- Multiple Mars landers, with the first lander(s) pre-deploying cargo to prepare for a later crew landing;
- Modest initial surface infrastructure: minimal surface power and communications infrastructure, but no surface habitat; and

- "All-up mission" approach: crew depart Earth with all the transit propellant they need for the round-trip journey, meaning no propellant will be harvested on Mars for the crew ascent vehicle for early missions.

These assumptions provide a framework for direct architecture comparisons, and all decisions will be made with evolution in mind. More complex mission scenarios will be addressed in subsequent analysis cycles, but the initial step is to define a practical approach to achieve the first human missions to Mars.

The primary Mars systems under study include the transportation system that will carry crew from an orbit in the Earth-Moon neighborhood and back; the entry, descent, landing, and ascent system that will land astronauts on the surface and return them to Martian orbit; the surface systems that will keep crew healthy and productive, allowing them to explore and conduct high-value science on the Red Planet; and the crew support systems that will provide the required logistics, food, and health care throughout their round-trip journey.

Chapter 7

NASA Appropriations and Authorizations: Fact Sheet*

Rachel Lindbergh

Overview

Congressional deliberations about the National Aeronautics and Space Administration (NASA) often focus on the availability of funding. This fact sheet provides data on past and current NASAappropriations, as well as the President's FY2025 budget request and congressional action on FY2025 appropriations and authorization of appropriations.

Table 1 shows budget authority for NASA for FY2019-FY2024. Except where noted, theamounts shown include regular appropriations, supplemental appropriations, rescissions,

transfers, and reprogramming. They are taken from NASA's congressional budget justificationsfor FY2021-FY2025;[1] the Consolidated Appropriations Act, 2024 (P.L. 118-42); and the explanatory statement for that act, *Congressional Record*, March 5, 2024, pp. S1141-S1142. Congressional budget justifications are available on the NASA budget website, https://www.nasa.gov/budgets-plans-and-reports/, for the current year and for past years back toFY2010.

---

* This is an edited, reformatted and augmented version of Congressional Research Service Publication No. R43419, updated August 7, 2024.

[1] FY2019 STEM Engagement amounts are not shown in the FY2021 congressional budget justification and are insteadtaken from the explanatory statement for the Consolidated Appropriations Act, 2020 (P.L. 116-93), *Congressional Record*, December 17, 2019, pp. H10969-H10971.

In: The National Aeronautics and Space Administration's Role …
Editor: Brent M. Jenkins
ISBN: 979-8-89530-539-3

Table 2 shows FY2024 appropriations provided by the Consolidated Appropriations Act, 2024 (P.L. 118-42), compared with FY2023 regular appropriations as enacted; the Administration's request for FY2024; FY2024 appropriations as recommended in the bill introduced by the chairof the House Appropriations Committee, Subcommittee on Commerce, Justice, Science, and Related Agencies (H.R. 5893 as introduced and explanatory material on the committee website at https://appropriations.house.gov/sites/republicans.appropriations.house.gov/files/FY24-CJS-Explanatory-Materials.pdf); and FY2024 appropriations recommended by the Senate Committee on Appropriations (S. 2321 as reported and S.Rept. 118-62). Note that the NASA Authorization Act of 2022 (P.L. 117-167, Title VII), which is the most recently enacted authorization, did not include authorizations of appropriations for NASA for FY2024 and beyond.

Table 3 compares FY2024 regular appropriations, as enacted, with the Administration's request for FY2025; FY2025 appropriations as recommended by the House Appropriations Committee, Subcommittee on Commerce, Justice, Science, and Related Agencies (H.R. 9026 as reported and H.Rept. 118-582); FY2025 appropriations as recommended by the Senate Committee on Appropriations (S. 2321 as reported and S.Rept. 118-62); and the FY2025 amounts that would beauthorized by H.R. 8958, referred to herein as the NASA Reauthorization Act of 2024, as orderedreported by the House Committee on Science, Space, and Technology on July 10, 2024.[2]

Additional columns will be added to this table as Congress acts to complete FY2025authorization and appropriations legislation for NASA.

A note on name changes: Since the FY2019 budget request, the Trump and Biden Administrationshave proposed renaming the Exploration account as "Deep Space Exploration Systems"; enacted appropriations have so far retained the name Exploration. Starting with the FY2023 budget, NASA renamed Exploration Systems Development as "Common Exploration Systems

Development," and Exploration Research and Development (R&D) as "Artemis Campaign Development." In the FY2025 budget request, NASA renamed Common Exploration Systems Development as "Moon to Mars

[2] H.R. 8958 was ordered to be reported but had not been reported as of August 7, 2024. While the official reported textof H.R. 8958 is not available, the amendments adopted during the markup did not appear to change the introduced bill provisions regarding FY2025 authorization amounts. (House Committee on Science, Space, and Technology, "Full Committee Markup of the NASA Reauthorization Act of 2024," July 10, 2024, https://science.house.gov/2024/7/full- committee-markup-of-the-nasa-reauthorization-act-of-2024.)

Transportation System," and Artemis Campaign Development as"Moon to Mars Lunar Systems Development."

Figure 1 shows NASA's total annual budget authority from the agency's establishment inFY1958 to FY2024, in both current dollars and inflation-adjusted FY2024 dollars.

Figure 1 shows NASA's total annual budget authority from the agency's establishment in FY1958 to FY2024, in both current dollars and inflation-adjusted FY2024 dollars.

Table 1. NASA budget authority, FY2019-FY2024

(in $ millions)

	FY2019	*FY2020*	*FY2021*	*FY2022*	*FY2023*	*FY2024*
Science	*$6,887*	*$7,143*[a]	*$7,291*	*$7,611*	*$7,792*	*$7,334*
Earth Science	1,931	1,972	1,997	2,061	2,175	2,195
Planetary Science	2,747	2,713	2,693	3,120	3,217	2,717
Astrophysics	1,191	1,306	1,356	1,394	1,510	1,530
James Webb Space Telescope	305	423	415	175	—[b]	—[b]
Heliophysics	713	725	751	778	805	805
Biological and Physical Sciences[c]	—	5	79	83	85	88
Aeronautics	*725*	*784*	*829*	*881*	*935*	*935*
Space Technology	*927*	*1,100*	*1,100*	*1,100*	*1,193*	*1,100*
Exploration	*5,045*	*5,960*	*6,397*	*6,855*	*7,448*	*7,666*[d]
Exploration Systems Development	4,087	4,513	4,539	4,591	4,717	4,533
Orion	*1,350*	*1,407*	*1,404*	*1,402*	*1,315*	*1,139*
Space Launch System	*2,144*	*2,528*	*2,555*	*2,600*	*2,567*	*2,600*
Exploration Ground Systems	*593*	*578*	*580*	*589*	*835*	*794*
Exploration R&D/Artemis Campaign Development	958	1,447	1,858	2,077	2,631	n/s
Other	—	—	—	187	101	n/s
Space Operations	*4,640*	*4,135*	*4,102*	*3,975*	*4,267*	*4,220*[e]
International Space Station	1,490	1,516	1,322	1,262	1,286	n/s
Space Transportation	2,110	1,746	1,872	1,717	1,760	n/s
Crew and Cargo Program	*1,896*	*1,511*	*1,573*	*1,570*	*1,642*	*1,856*
Space and Flight Support	1,000	857	890	889	983	n/s
Commercial LEO Development	40	15	18	102	224	228
Exploration Operations	—	—	—	5	13	n/s
STEM Engagement	*110*	*120*	*127*	*137*	*144*	*143*
Space Grant	44	48	51	55	58	58
EPSCoR	21	24	26	26	26	26
MUREP	33	36	38	43	46	46

Table 1. (Continued)

	FY2019	*FY2020*	*FY2021*	*FY2022*	*FY2023*	*FY2024*
Other	12	12	12	14	14	14
Safety, Security, & Mission Services	2,755	2,913[f]	2,937	3,021	3,137	3,129
Construction and EC&R	372	433	446	417	612g	300
Inspector General	39	42	44	45	48	48
Total	21,500	22,629[a,f]	23,271	24,041	25,573g	24,875

Sources: FY2019-FY2023 from NASA FY2021-FY2025 congressional budget justifications. FY2024 from P.L. 118-42 and explanatory statement, Congressional Record, March 5, 2024, pp. S1141-S1142.

Notes: Except where noted, amounts include regular appropriations, supplemental appropriations, rescissions, transfers, and reprogramming. Figures may not sum to totals because of rounding. R&D = Research and Development; LEO = Low Earth Orbit; STEM = Science, Technology, Engineering, and Mathematics; EPSCoR = Established Program to Stimulate Competitive Research; MUREP = Minority University Research and Education Program; EC&R = Environmental Compliance and Remediation; n/s = not specified.

[a] Not adjusted to reflect rescission of $70 million from prior year unobligated balances (Section 521(c)).

[b] Included in Astrophysics.

[c] Included in International Space Station before FY2020.

[d] Includes $3.133 billion appropriated to the Exploration account by P.L. 118-42 without specified purpose.

[e] Includes $2.136 billion appropriated to the Space Operations account by P.L. 118-42 without specified purpose.

[f] Does not include additional $60 million appropriated by the CARES Act (P.L. 116-136).

[g] Includes $556.4 million in emergency supplemental funding provided in Division N of P.L. 117-328.

Table 2. NASA appropriations, FY2024

(budget authority in $ millions)

FY2024 Appropriations					
	FY2023 Enacted	Request	House Intro.	Senate Cmte.	Enacted
Science	$7,795	$8,261	$7,380	$7,341	$7,334
Earth Science	2,195	2,473	2,000	2,219	2,195
Planetary Science	3,200	3,383	3,100	2,683	2,717
Astrophysics	1,510	1,557	1,485	1,544	1,530
Heliophysics	805	751	710	805	805
Biological and Physical Sciences	85	97	85	90	88
Aeronautics	935	996	946	935	935
Space Technology	1,200	1,392	1,205	1,118	1,100
Exploration/Deep Space Exploration Systems	7,469	7,971	7,971	7,736	7,666
Common Exploration Systems Development	4,738	4,525	4,525	4,525	4,533
Orion	*1,339*	*1,225*	*1,225*	*1,225*	*1,139*
Space Launch System	*2,600*	*2,506*	*2,506*	*2,506*	*2,600*

FY2024 Appropriations					
	FY2023 Enacted	Request	House Intro.	Senate Cmte.	Enacted
Exploration Ground Systems	*799*	*794*	*794*	*794*	*794*
Artemis Campaign Development	2,600	3,235	3,235	n/s	n/s
Human Exploration Requirements and Architecture	n/s	49	n/s	n/s	n/s
Mars Campaign Development	n/s	162	n/s	n/s	n/s
Space Operations	4,250	4,535	4,345	4,200	4,220
International Space Station	n/s	1,303	n/s	n/s	n/s
Space Transportation	n/s	1,957	n/s	n/s	n/s
Crew and Cargo Program	*n/s*	*1,856*	*n/s*	*1,856*	*1,856*
Space and Flight Support	n/s	1,047	n/s	n/s	n/s
Commercial LEO Development	224	228	n/s	228	228
STEM Engagement	144	158	89	144	143
Space Grant	58	58	60	58	58
EPSCoR	26	26	29	26	26
MUREP	46	48	0	46	46
Other	14	26	0	14	14
Safety, Security, and Mission Services	3,129	3,369	3,136	3,100	3,129
Construction and EC&R	604[a]	454	248	379	300
Inspector General	48	50	48	48	48
Total	25,573[a]	27,185	25,367	25,000	24,875

Sources: FY2023 Enacted: P.L. 117-328 and explanatory statement, Congressional Record, December 20, 2022, pp. S7945-S7950. FY2024 Request: FY2024 NASA congressional budget justification. FY2024 House Introduced:

H.R. 5893 as introduced and explanatory material on the House Appropriations Committee website at https://appropriations.house.gov/sites/republicans.appropriations.house.gov/files/FY24-CJS-Explanatory- Materials.pdf. FY2024 Senate Committee: S. 2321 as reported and S.Rept. 118-62. FY2024 Enacted: P.L. 118-42 and explanatory statement, Congressional Record, March 5, 2024, pp. S1141-S1142.

Notes: Figures may not sum to totals due to rounding. LEO = Low Earth Orbit; STEM = Science, Technology, Engineering, and Mathematics; EPSCoR = Established Program to Stimulate Competitive Research; MUREP = Minority University Research and Education Program; EC&R = Environmental Compliance and Remediation; n/s

= not specified.

[a] Includes $556.4 million in emergency supplemental funding provided in Division N of P.L. 117-328.

Table 3. NASA appropriations, FY2025, and proposed FY2025 authorization

(budget authority in $ millions)

FY2025 Appropriations						FY2025 Auth.
	FY2024 Enacted	Request	House Cmte.	Senate Cmte.	Enacted	House Cmte[a]
Science	$7,334	$7,566	$7,334	$7,576		$7,334
Earth Science	2,195	2,379	2,000	2,369		n/s

Table 3. (Continued)

FY2025 Appropriations						FY2025 Auth.
	FY2024 Enacted	Request	House Cmte.	Senate Cmte.	Enacted	House Cmte[a]
Planetary Science	2,717	2,732	2,930	2,722		n/s
Astrophysics	1,530	1,578	1,532	1,583		n/s
Heliophysics	805	787	787	812		n/s
Biological and Physical Sciences	88	91	85	91		n/s
Aeronautics	935	966	966	966		966
Space Technology	1,100	1,182	1,182	1,182		1,182
Exploration/Deep Space Exploration Systems	7,666	7,618	7,618	7,648		7,618
Common Exploration Systems Development[b]	4,533	4,213	4,738	4,213		n/s
Orion	*1,139*	*1,031*	*1,339*	*1,031*		n/s
Space Launch System	*2,600*	*2,423*	*2,600*	*2,423*		n/s
Exploration Ground Systems	*794*	*759*	*799*	*759*		n/s
Artemis Campaign Development[c]	n/s	3,288	n/s	n/s		n/s
Human Exploration Requirements and Architecture	n/s	117	n/s	n/s		n/s
Mars Campaign Development[d]	n/s	—	n/s	n/s		n/s
Space Operations	4,220	4,390	4,474	4,400		4,474
International Space Station	n/s	1,270	n/s	n/s		n/s
Space Transportation	n/s	1,862	n/s	n/s		n/s
Crew and Cargo Program	1,856	1,762	1,890	n/s		n/s
Space and Flight Support	n/s	1,088	n/s	n/s		n/s
Commercial LEO Development	228	170	n/s	170		n/s
STEM Engagement	143	144	89	144		135
Space Grant	58	57	60	59		n/s
EPSCoR	26	25	29	26		n/s
MUREP	46	46	0	46		n/s
Other	14	15	0	14		n/s
Safety, Security, and Mission Services	3,129	3,044	3,044	3,044		3,044
Construction and EC&R	300	424	424	424		424
Inspector General	48	51	48	51		48
Total	24,875	25,384	25,179	25,245		25,225

Sources: FY2024 Enacted: P.L. 118-42 and explanatory statement, Congressional Record, March 5, 2024, pp. S1141-S1142. FY2025 Request: FY2025 NASA congressional budget justification. FY2025 House: appropriations recommended by the House Appropriations Committee (H.R. 9026 as reported and H.Rept. 118-582). FY2025 Senate: appropriations recommended by the

Senate Appropriations Committee (S. 4795 as reported and S.Rept. 118-198). FY2025 House Authorization: NASA Authorization Act of 2024 (H.R. 8958), as ordered reported by the House Committee on Science, Space, and Technology on July 10, 2025, https://science.house.gov/markups?ContentRecord_id=F6543C82-0980-435B-BFDF-4DC3EF41EFD5.

Notes: Figures may not sum to totals because of rounding. LEO = Low Earth Orbit; STEM = Science, Technology, Engineering, and Mathematics; EPSCoR = Established Program to Stimulate Competitive Research; MUREP = Minority University Research and Education Program; EC&R = Environmental Compliance and Remediation; n/s = not specified.

[a] H.R. 8958 was ordered to be reported but had not been reported as of August 7, 2024. While the official reported text of H.R. 8958 is not available, the amendments adopted during the markup did not appear to change the introduced bill provisions regarding FY2025 authorization amounts.

[b] Referred to in the FY2025 request as "Moon to Mars Transportation System."

[c] Referred to in the FY2025 request as "Moon to Mars Lunar Systems Development."

[d] As noted in the FY2025 request, NASA intends to retire its Mars Campaign Development theme and will reallocate its content elsewhere in the Exploration Systems Development Mission Directorate (p. DEXP-69).

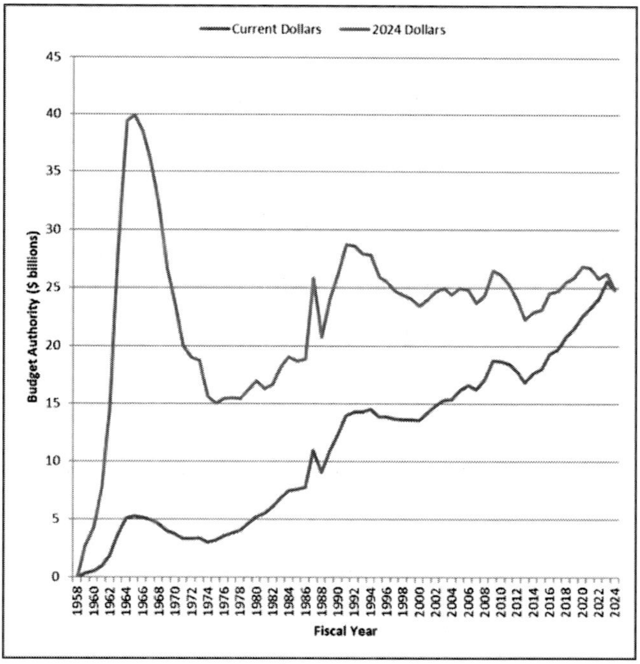

Sources: Compiled by CRS. FY1958-FY2008 from NASA, Aeronautics and Space Report of the President: Fiscal Year 2008 Activities, Table D-1A, https://ntrs.nasa.gov/citations/20110012306. FY2009-FY2023 from NASA congressional budget justifications. FY2011-FY2025, adjusted for supplemental appropriations, rescissions, and sequestration not shown in the justifications. FY2024 from P.L. 118-42. Current dollars deflated to FY2024 dollars using gross domestic product (GDP) (chained) price index from President's budget for FY2025, Historical Table 10.1, https://www.whitehouse.gov/omb/historical-tables/.

Note: Transition quarter between FY1976 and FY1977 not shown.

Figure 1. NASA Funding, FY1958-FY2024.

Index

A

Africa, 63, 72, 106, 202
Appropriations Act, 25, 26, 55
artificial intelligence, 160, 169, 189
asteroids, 11, 24, 74, 123, 166, 225

B

Beijing, 32, 40
biotechnology, 104
broadband, 182
building blocks, 74, 218

C

capital markets, 36, 41, 54, 55
China, 21, 25, 30, 32, 34, 37, 40, 41, 43, 44, 45, 46, 47, 48, 51, 52, 54, 56, 63, 64, 66, 69, 70, 71, 72, 73, 74, 75, 82, 85, 87, 93, 95, 104, 105, 106, 107, 111, 114, 115, 120, 121, 122, 137, 140, 155, 165, 176, 177, 190
civil space exploration, 200
climate change, 115, 121, 123, 163
coal, 164, 166, 167, 187, 193
cold war, 28, 30, 31
communication, 92, 182
computer systems, 165, 168
computing, 157, 160, 189, 195
concurrent engineering, 138, 144
cybersecurity, 102

D

dark energy, 225
data analysis, 13
database, 26
demographic factors, 7
Department of Defense, 23, 24, 71, 73, 80, 160
Department of Energy, 96, 97
developing countries, 63, 104, 106
developing nations, 72
differential equations, 193
distillation, 238
domestic industry, 173
dominance, 40, 46, 49, 114, 174
drought, 11, 175

E

economic development, 42, 54, 57, 197, 200
economic downturn, 152, 156
ecosystem, 80, 96, 98, 99
education, vii, 1, 2, 3, 4, 10, 19, 20, 37, 42, 44, 54, 55, 160, 189
electricity, 15
encouragement, 55, 82, 113, 120
engineering, vii, 1, 3, 5, 9, 26, 51, 69, 77, 78, 79, 80, 85, 109, 112, 116, 136, 142, 145, 163, 171, 201, 228
entrepreneurial sector, 48, 153, 158
environment, 39, 63, 64, 68, 69, 76, 77, 78, 80, 93, 100, 101, 104, 107, 117, 124, 125, 132, 137, 141, 144, 156, 196, 206
environmental impact, 19, 50
environmental sustainability, 73
European Union, 15, 34
exploitation, 41

F

FAA, 89, 90, 91, 95, 118, 119, 120, 126, 134
fault tolerance, 138, 143
Federal Communications Commission, 24, 80
fission, 213
foreign affairs, 74
fuel cell, 141

H

human rights, 25
hydrogen, 205

I

India, 73, 176, 177, 190
industrial policy, 48
international competitiveness, 69, 77, 80
International Space Station, 1, 3, 15, 16, 17, 28, 30, 31, 32, 35, 36, 39, 40, 43, 44, 63, 87, 88, 89, 104, 105, 106, 107, 108, 109, 116, 121, 125, 134, 135, 154, 156, 157, 162, 163, 167, 175, 202, 203, 222, 235, 245, 246, 247, 248

J

Japan, 16, 32, 44, 65, 72, 88, 120, 207, 233

K

Kennedy, John Fitzgerald, 145, 148, 149, 164, 166
Korea, 136, 139

M

macular degeneration, 105
mercury, 133, 136, 137, 140, 141, 142, 148, 150, 181, 187, 188
Mexico, 96, 98
microgravity, 13, 35, 36, 63, 68, 69, 75, 97, 104, 105, 107, 111, 175

N

National Aeronautics and Space Administration, vii, 1, 3, 8, 25, 45, 56, 78, 112, 120, 124, 199, 217, 243
National Defense Authorization Act, 49, 114, 115
National Oceanic and Atmospheric Administration (NOAA), 15, 80, 114, 118
National Research Council, 44, 45, 134
National Security Council, 56
national strategy, 47, 56
natural disasters, 11

O

Obama, President Barack, 3
organizational culture, 114, 118

P

parallel, 165, 168, 169
participants, 44, 119, 125, 162, 175
pathways, 14, 46, 150, 161
Pentagon, 135
physical sciences, 13
planets, 11, 12, 130
platform, 18, 57, 121, 205, 206, 207, 208, 209, 210, 212, 213, 222
policy, 22, 24, 38, 42, 53, 54, 55, 56, 57, 62, 68, 70, 71, 72, 74, 83, 92, 103, 112, 118, 119, 124, 152, 156, 157, 158, 168, 200
policy issues, 92, 124
political system, 66, 154
portfolio, 12, 41, 50, 53, 157, 173, 175
power generation, 240
preparation, 32, 148, 160, 175, 181, 200, 203, 204, 207, 210
principles, 24, 44, 45, 51, 63, 74, 75, 79, 182, 183, 200
private sector, 3, 15, 30, 32, 54, 63, 64, 67, 69, 75, 85, 88, 89, 91, 92, 93, 95, 97, 98, 100, 123, 124, 171, 174, 184

procurement, 22, 24, 41, 47, 53, 158, 159, 178, 183, 184
professionals, 117, 151, 159, 160, 186, 201
project, 20, 30, 47, 54, 89, 142, 176, 188
protection, 51, 208
prototype, 154, 162, 238
public safety, 118, 119, 123, 127
public-private partnerships, 57, 64, 71, 92, 97, 98, 117, 172, 180
purchasing power, 170, 173

R

race, 32, 70, 94, 102, 152, 155, 183
radiation, 205, 208, 209, 214
radio, 80, 200, 213
reality, 28, 40, 43, 68, 132
recycling, 110
regenerative medicine, 105
regulations, 24, 77, 80, 196
regulatory framework, 77, 80, 123, 125, 158
regulatory system, 100
remote sensing, 80, 175, 182
renaissance, 171, 174
researchers, 40, 43, 51, 57, 239
resilience, 70, 136, 139
resources, 13, 20, 21, 45, 54, 63, 77, 81, 83, 101, 120, 125, 126, 127, 128, 137, 140, 141, 158, 163, 183, 196, 200, 204, 205, 208, 209, 239, 240
robotics, 37, 64, 69, 70, 73, 104, 121, 192, 207, 213, 228
rules, 37, 71, 73, 74, 75, 80, 96, 99, 136, 138, 140, 196
Russia, 15, 16, 30, 34, 37, 44, 46, 51, 56, 87, 88, 93, 95, 156, 176, 207

S

safety, 16, 19, 39, 76, 77, 78, 80, 81, 83, 88, 90, 98, 108, 118, 119, 120, 123, 124, 128, 135, 138, 143, 144, 148, 150, 153, 177, 179
satellite service, 100
Saturn, 148, 150, 181, 218
science, 2, vii, 1, 2, 3, 5, 6, 9, 10, 11, 12, 14, 15, 16, 18, 20, 29, 31, 35, 37, 41, 42, 44, 46, 48, 50, 51, 54, 75, 83, 97, 109, 116, 120, 122, 126, 151, 153, 155, 157, 159, 161, 162, 163, 164, 169, 171, 176, 182, 189, 191, 192, 193, 201, 204, 203, 213, 221, 227, 229, 230, 233, 235, 237, 239, 241, 244, 249
scientific investigations, 182, 205, 206, 212
scientific understanding, 200
security, 25, 30, 37, 49, 55, 63, 69, 70, 73, 77, 80, 83, 95, 115, 119, 132, 152, 155, 159, 160, 173
sensors, 11, 69
services, 1, 2, 3, 9, 17, 41, 42, 50, 51, 52, 53, 54, 55, 57, 76, 78, 79, 104, 113, 115, 174, 184, 235, 236, 239
simulation(s), 141, 155, 179, 239
software, 128, 206
solar system, 161, 162, 163, 199, 200, 205, 206, 218
solution, 89, 91, 94, 97, 171
Soviet Union, 131, 137, 140, 154, 177
Soyuz, 2, 16
space environment, 64, 123, 124, 125, 206
space shuttle, 7, 16, 167
space station, 2, 17, 32, 35, 36, 40, 42, 51, 63, 68, 69, 82, 85, 95, 100, 104, 105, 106, 107, 108, 109, 110, 111, 113, 114, 121, 122, 123, 125, 155, 221, 222, 231
space technology, vii, 1, 3, 10, 18, 55, 70, 75
stability, 44, 56, 83, 112, 158, 178, 196
stars, 12
structure, 65, 77, 80, 86, 90, 115, 142, 178, 179
suppliers, 55, 113, 189
supply chain, 49, 233
surface area, 202
surveillance, 48, 99
sustainability, 28, 149, 151, 158, 192
sustainable development, 212

T

techniques, 143, 165, 168, 204
technological advances, 90, 133
tension(s), 16, 30, 44
territory, 38, 102, 105
testing, 47, 84, 85, 95, 97, 138, 143, 181, 182, 206, 211, 213
thoughts, 40, 42, 152, 154, 196
threats, 55, 71
titanium, 166
Title I, 20
Title II, 20
Title V, 22, 244
trade, 42, 96, 105, 135, 160, 173
trade agreement, 105
training, 16, 117, 122, 132, 137, 141, 143, 148, 150, 155
trajectory, 143, 231
transition period, 56, 107
transparency, 63, 73, 88, 159, 237
transport, 1, 2, 3, 16, 209, 233, 234
transportation, 41, 115, 118, 119, 128, 153, 162, 177, 180, 206, 222, 232, 239, 241

U

U.S. Geological Survey (USGS), 11, 132
Ukraine, 16
Ultra High Energy Cosmic Rays, 2
United Nations, 25, 52, 74, 95, 123
United States, 15, 16, 17, 25, 27, 30, 36, 37, 43, 44, 45, 46, 47, 49, 51, 80, 82, 83, 94, 95, 96, 97, 98, 99, 100, 102, 103, 104, 112, 114, 116, 119, 121, 125, 127, 129, 131, 132, 133, 135, 137, 140, 152, 155, 156, 159, 161, 163, 164, 165, 168, 171, 172, 173, 174, 177, 181, 183, 184, 189, 194, 195, 199, 200, 201, 207, 227
universe, 12, 116, 117, 130, 151, 165, 169, 200, 205, 208, 218
universities, 20, 96, 116, 117, 121, 122, 125, 126, 133, 190, 206
uranium, 168
urban, 11

V

vacuum, 62, 130
vehicles, 97, 125, 153, 162, 174, 176, 178, 182, 203, 220
venture capital, 49
Vietnam, 134, 137, 140, 165
vision, 33, 65, 67, 72, 79, 131, 148, 150, 170, 185, 195
visualization, 41, 224

W

war, 16, 95, 103, 154
Washington, 27, 33, 45, 84, 129, 134, 135, 165, 167, 183, 192, 194
waste, 161, 209
waste disposal, 209
water, 163, 165, 169, 200, 205, 212, 213
witnesses, 29, 32, 84, 88, 92, 107, 108, 111, 118, 122, 126, 130, 133, 184, 197
workforce, 7, 19, 122, 133, 141, 148, 159, 160, 161, 187, 188, 189, 190, 191, 204
World War I, 139, 142, 181

Y

yield, 31, 35, 107, 133
young people, 166, 185, 191, 194
young women, 191